The Calling of Global Responsibility

I0042021

This book rethinks and transforms the current discourse on globalization and global justice. It expands the idea of globalization from an economic or corporate context to mean humanization and planetary realizations— moving beyond the boundaries of nation-states and other human-made demarcations. The author challenges the notion of human primacy and makes a fervent call to reconfigure the paradigm of anthropocentrism. Through a careful study of movements for justice and inter-faith dialogue from across the world, the book makes a unique contribution to the emerging study of global responsibility. It also helps us overcome our current civilizational crises and cultivate a new civilization of planetary care and co-responsibility.

As part of the *Ethics, Human Rights and Global Political Thought* series, the volume will be of great interest to scholars and researchers of law and society, especially social movements, political theory, and philosophy.

Ananta Kumar Giri is Professor at the Madras Institute of Development Studies, Chennai, India.

Ethics, Human Rights, and Global Political Thought

Series Editors: Aakash Singh Rathore and Sebastiano Maffettone
Center for Ethics & Global Politics, Luiss University, Rome

Whereas the interrelation of ethics and political thought has been recognized since the dawn of political reflection, over the last sixty years—roughly since the United Nation's Universal Declaration of Human Rights—we have witnessed a particularly turbulent process of globalizing the coverage and application of that interrelation. At the very instant the decolonized globe consolidated the universality of the sovereign nation-state, that sovereignty—and the political thought that grounded it—was eroded and outstripped, not as in eras past, by imperial conquest and instruments of war, but rather by instruments of peace (charters, declarations, treaties, conventions), and instruments of commerce and communication (multinational enterprises, international media, global aviation and transport, internet technologies).

Has political theory kept a pace with global political realities? Can ethical reflection illuminate the murky challenges of real global politics?

This Routledge book series *Ethics, Human Rights and Global Political Thought* addresses these crucial questions by bringing together outstanding monographs and anthologies that deal with the intersection of normative theorizing and political realities with a global focus. Treating diverse topics by means of interdisciplinary techniques—including philosophy, political theory, international relations and human rights theories, and global and postcolonial studies—the books in the Series present up to date research that is accessible, practical, yet scholarly.

Civil Disobedience from Nepal to Norway
Traditions, Extensions, and Civility
Edited By Tapio Nykänen, Tiina Seppälä, Petri Koikkalainen

The Calling of Global Responsibility
New Initiatives in Justice, Dialogues and Planetary Realizations
Ananta Kumar Giri

For more information about this series, please visit: www.routledge.com/ Ethics-Human-Rights-and-Global-Political-Thought/book-series/EHRGPT

The Calling of Global Responsibility

New Initiatives in Justice, Dialogues and Planetary Realizations

Ananta Kumar Giri

Routledge
Taylor & Francis Group

LONDON AND NEW YORK

First published 2023
by Routledge
4 Park Square, Milton Park, Abingdon, Oxon OX14 4RN

and by Routledge
605 Third Avenue, New York, NY 10158

Routledge is an imprint of the Taylor & Francis Group, an informa business

© 2023 Ananta Kumar Giri

The right of Ananta Kumar Giri to be identified as author of this
work has been asserted in accordance with sections 77 and 78 of the
Copyright, Designs and Patents Act 1988.

British Library Cataloguing-in-Publication Data
A catalogue record for this book is available from the British Library

Library of Congress Cataloging-in-Publication Data
A catalog record for this book has been requested

ISBN: 978-0-367-36503-5(hbk)
ISBN: 978-1-032-44186-3(pbk)
ISBN: 978-0-429-34748-1(ebk)

DOI: 10.4324/9780429347481

Typeset in Sabon
by Apex CoVantage, LLC

For Philip Quarles van Ufford
Dipesh Chakraborty
Jill-Carr Harriss
Enrique Dussel
Pradjarta
Sailesh Rao
Sudha Sreenivas Reddy

Contents

Global Responsibility and the Calling of Planetary
Realizations: Walking and Meditating Together With
Epilogues and Reflections   248
ANANTA KUMAR GIRI

Foreword

In this book, the author explores new initiatives which today are opening up possibilities towards the cultural and social embodiment of global responsibility and thus contributing, whether intentionally or effectively, to the emergence of this urgently needed collective good. This innovative exploration traverses a number of interrelated dimensions, from individual social movements, via dialogues between different sociocultural groupings and the internal transformation of corporations, to the critical problems and challenges of our contemporary polycrisian age and the potential planetary outcomes of intercultural communication and collective activities. Reinforcing the author's undertaking, the volume is closed by a roundtable of reflections on the theme of the book by 12 authors from around the globe. Given that the notion of responsibility is nothing less than a civilizational achievement and cultural asset of humanity, the author is able to selectively link up with certain themes from its history, to creatively continue or vary them and to make imaginative yet plausible projections for the future. Of particular importance, however, is the focal concept of global responsibility—a conception that emerged in two different senses over a period of more than 40 years due to developments in society, on the one hand, and the genesis of movements in response to those changes, on the other hand.

In the first instance, the conception of global responsibility arose in conjunction with the mid-20th-century problem of the threatening consequences of industrial-scale technological development. In the wake of the devastating consequences of the bombing of Japan and accidents at nuclear power plants in the United Kingdom and the USA, the 1950s and 1960s witnessed not only a growing awareness of the spectrum of low-probability yet high-consequence risks for both the environment and future generations, but also the concomitant formation of the peace and environmental branches of a broad new social movement. These developments laid the basis for the full-scale problematization of the internal organization of contemporary crisis-ridden society and its relation to nature as well as the diverse yet interrelated movements which are currently present in all their vigour. Simultaneously, they also provided the context for the reflexive abstraction and formulation of the core issue implicated: who should bear responsibility for the catalogue of problems and their risk-laden consequences? The first reaction came from scientists from a variety of disciplines who entered a self-reflective debate about their responsibility for contributing to past problems and their consequences.

By the late 1970s and early 1980s, the answer to the question of the subject of responsibility was forthcoming. Hans Jonas coined the new concept of "collective responsibility," while Karl-Otto Apel refined it as the "universalistic macro-ethics of co-responsibility" so as to accommodate the insight that the global collective shouldering of this burden by no means excludes individuals, but rather demands that each and every one of us take the necessary responsibility. Such responsibility includes the obligations to take one's rights seriously and exercise them by sharpening one's sensitivity to problems and risks, nurturing one's competence to understand and articulate them, bolstering one's courage and expressive power to expose and publicize them, and participating in activities bearing on related decision-making. Simultaneously it was clear that, in doing so, concomitant considerations of truth, right, justice, freedom, equality, solidarity, legitimacy, truthfulness, dignity, and appropriateness needed to be observed in all relevant evaluations and decisions.

In the second instance, the conception of global responsibility received another or, perhaps, an extended meaning in the context of the changed conditions of the 1990s. In the wake of such momentous events as the so-called "European Revolution" of 1989, including the dissolution of the Soviet Union, the fall of the Berlin Wall, and the liberation of civil society in Eastern Europe, as well as the significant advancement of decolonization with the end of Apartheid in South Africa in 1994, a new vision of the human sociocultural world emerged: a cosmopolitan world of which every individual is a citizen and in the global context of which the violation of rights in any one place is felt and condemned throughout. While traceable back at least as far as Greco-Roman antiquity and the European Enlightenment and its influence sphere, this conception of the cosmopolitan condition found significant affirmation in 1995. This was the year of the multiple anniversaries of Immanuel Kant's proposal for perpetual peace from a cosmopolitan perspective, the end of World War II and the establishment of the United Nation's Charter. As in the earlier case, the emergence of a broad and vigorous cosmopolitan movement accompanied these various developments. Its diverse representatives, including individual world citizens, philosophers, social scientists, historians, and legal scholars as well as organizations and agencies like the United Nations, the International Law Commission, and NGOs are not only offering theoretical clarifications of what is required, but are also committed practically through the pursuit of particular goals to giving sociocultural and hopefully also institutional effect to it. Central to these efforts is the acknowledgement of the moral obligation—in the sense of global cosmopolitan co-responsibility—that we as members of a generation in a chain of historical and future generations have to contribute towards the establishment of the conditions necessary for bringing the projected global cosmopolitan order into being for our children and grandchildren.

Already in the course of their development, the environmental and cosmopolitan conceptions of global responsibility tended to become interrelated. By the early 21st century, consequently, the convergence of the two senses or streams culminated in the broadened idea, stripped of every anthropocentric trace, of a global cosmopolitan existence of all forms of life on earth in an ecologically cared-for planetary bio-social ecosphere. It is this conception which forms the general background that the author assumes for the purposes of this book.

Rather than just a general background, however, what he more specifically assumes is the notion of responsibility as a world-disclosing, world-building, and world-ordering concept. As they have done with increasing intent over millennia since the so-called "Human Revolution" and the subsequent "cultural explosion" of some 40,000 years ago, humans construct their own unique sociocultural world by means of the constellation of fundamental concepts to which they themselves have given rise in the course of time. This constellation of concepts, among which is also responsibility, has absorbed the power of their onomatopoeic and poetic origin, crystallized out and become more and more precisely articulated, as is, for example, attested by the primary objective, social, and subjective sets coined in ancient Indian (substance, morality, and happiness), ancient Greece (truth, the good, and beauty), and in modern society (truth, right/justice, and truthfulness/ appropriateness). It is by means of such primary concepts and those falling under them in the theoretical-empirical, moral, and ethical formal-ontological categories that humans in the past have created, organized, transformed, and reproduced their sociocultural worlds and, today, are in the process of constructing the global cosmopolitan world in which all forms of life can exist in an ecologically cared-for planetary bio-social ecosphere. The deep-seated understanding which runs like an unbroken thread through this entire book is that the world-disclosing, world-building, and world-ordering normative concept of global responsibility, in conjunction with its surrounding concepts, is absolutely central to the contemporary endeavour to contribute meaningfully to the establishment of the conditions for the nascent world to make its definitive appearance.

Against this general background, this book is located in the historical, social, cultural, economic, and political context of the past 30 years or so and the multi-dimensional crisis of our time. In line with what may be called its conceptual realist approach, its aim is to probe new initiatives and departures in the human endeavour to practically construct a global cosmopolitan and ecologically sustainable planetary sociocultural world in a way which answers to the very world-defining concepts that over millennia had emerged from social relations and practical engagements. The initiatives and departures singled out in the account all exhibit recourse in one way or another to the spectrum of these civilization-founding concepts, such as truth, rights, justice, freedom, equality, solidarity, and dignity. Above all, however, the author is on the lookout for how the concept of global responsibility comes, or is brought, into play as the most pertinent horizon calling—as announced by the title of the book—for observance in the contemporary epoch. As regards the sociocultural incarnation or embodiment of the concept of global responsibility, he stresses the necessity of a practical process dubbed "responsibilization."

As regards the initiatives and departures probed for their actual or potential contribution to the ongoing process of the responsibilization, the net is rightly cast to capture a wide range of phenomena of our time which are chapter by chapter subjected to analysis. While an extensive corpus of literature relative to these phenomena is surveyed, the approach adopted includes a geographically wide-ranging ethnographic exploration and, where possible, interviews and discussions with those driving and involved in them.

The study of social movements, for example, profiles the Indian Ekta Parishad movement that seeks to advance the interests of the landless and roofless, the comparable Brazilian Landless Workers' Movement, the Belgian-based transnational agriculture, food, and global justice movement Via Campesina, and the European anti-neoliberal and tax justice movement ATTAC that generalizes its concerns to the global justice level. Still other movements which receive variable degrees of coverage are the transnational World Social Forum, the Spanish Indignados, and the US Occupy movement. The struggles of these social movements are all understood as displaying attempts at the practical embodiment of such concepts—rights, justice, and dignity—so as to bring into being and organize the emerging society. Although pursuing their own selected values rather than explicitly seeking to give embodiment to global responsibility as such, the movements are nevertheless considered as effectively being vehicles of the process of responsibilization.

The same concern motivates the probes conducted into the remaining phenomena. The foray into the field of attempts to transcend confining particularisms consists of probes of both more particular inter-religious and more general inter-civilizational boundary-transgressing dialogues. Focused on inter-religious encounters in Indonesia, explorations are undertaken of a variety of initiatives by politicians, theologians, research and educational institutions, and voluntary organizations to assuage conflicts, fears, and tensions between organized religions such as Islam, the major branches of Christianity and Buddhism, and indigenous spirituality by facilitating communication between them at different levels. In addition to visits to organizations and interviews and discussions with individuals, the author himself on occasion ventures to contribute to this effort in bridge-building communication. Various different cases of inter-civilizational dialogues are presented, with attention given in particular to the efforts in and around the United Nations following the so-called "9/11" attack on the World Trade Center in New York. The initial response was the declaration of 2001 as the UN Year of Dialogue Among Civilizations, but concerted actions culminated a few years later in the UN Alliance of Civilizations. Overseen by a High-Level Group, it proceeded according to a formalized set of principles and a large body of multi-level programmes aimed at advancing inter-civilizational dialogue from the ground up and the top down. Highlighted here is the central thrust of this initiative which is the attainment of global justice and, thereby, the advancement also of global responsibility. As the United Nations' agenda more recently became filled by a range of other issues, various new initiatives that continue to pursue the aim of inter-civilizational dialogue are identified. Among them, for example, are the Dialogue of Humanity, founded in France in 2010 but now a global organization, and the Foundation for Universal Responsibility of His Holy the Dalai Lama which pursues a number of different programmes and also cooperates with a variety of affiliated organizations.

The presentation of boundary-transgressing dialogues in the book largely follows an ethnographic or participant-observational approach. But interesting is the complementary attention paid to the problem of intercultural communication, which allows the account to penetrate somewhat deeper into the problem. Of particular importance in the case of intercultural communication, considered social-psychologically and communication-theoretically, are the superficiality of the initial categorization in which groups confronting each other engage, followed by the drama of the encounter of their fixed identities and, finally, the not untypical failure on the part of those involved to appreciate that inter-cultural communication must start from the acknowledgement of ineradicable differences as well as of the presupposition of shared rules of communication. Whereas the attainment of such acknowledgement is the primary blockage to intercultural understanding, it is also the gateway to global responsibility. Initiatives towards boundary-transgressing dialogues and intercultural communication are thus of the essence as drivers of the process of responsibilization.

To underscore the contemporary relevance of the variety of initiatives and departures investigated in this book, the author both broadens his perspective and makes it more specific, historically and concretely. The social movements, boundary-transgressing dialogues, and intercultural communication as well as the various steps towards the construction of a global cosmopolitan and ecologically secured planetary society are contextualized by an acknowledgement and exploration of the critical problems and challenges of our contemporary polycrisian age. For this purpose, attention is given to the identification and isolation of the different components of the multiple overlapping crises of our time.

The first component is the bloated global corporate sector that has grown so powerful that it has turned out to be a major factor in both the generation of unprecedented inequality and far-reaching ecological damage. At a minimum, consequently, the sector's culture requires a radical transformation from corporate egoism towards social and ecological responsibility on a global scale. The second component is an attitude, epistemology, and corresponding set of practices that, rather than being ecologically attuned, favours a blind, unthinking, and exploitative relation to nature which has led to a degree of global warming and climate change that has brought humanity to the brink of an irreversible dynamic—what the IPCC report of August 9, 2021 signalled as "a code red for humanity." If global warming and climate change demand a root and branch transformation of our epistemological stance on the basis of the recognition that we are actually able to see the relations of everything to everything else, then the third component of the contemporary multi-dimensional crisis calls for an intensification of our appreciation of our own agency. Recently, under the disenchanting conditions of the new geological epoch of the Anthropocene, the epoch of the infringement of humans activities on nature, we came to the realization that, far from being distinct from nature, humanity is a part of and actually a geophysical and biomorphic force operating in nature which has grown so potent that it is capable of destroying its own organic foundations of life and thereby itself. Only a proper grasp of its own agency and a reflexive orientation towards global responsibility and a planetary-wide embodiment of this concept can save humanity from species suicide. The closely related COVID-19 health emergency is just the latest symptom of the multi-dimensional crisis of our time that reinforces the urgent need for us humans to own up to the devastating and self-injuring consequences of our frame of mind, mode of thinking, and typical practices and, correspondingly, to engage in the equally multi-level transformation required of us.

The central concern of this book and the differentiated way in which it is analysed and presented make it a timely contribution to the intertwined national and international debates about the nascent global society and the many challenges it faces. The actuality of this contribution and its critical but hopeful tone is all the more important currently in proportion as an autocratic political wave engaged in misguiding sophistry, and threatening to turn the clock back has in the meantime begun to sweep the world and, concomitantly, is impeding the ongoing construction of a planetary-wide global society. Particularly concerning is its abuse, hollowing out, and deriding of precisely the millennia-old formal-ontological conceptual conditions of the sociocultural world, including the concept of responsibility. What is certain, by contrast, is that the practical pursuit and realization envisaged in this book of the world-disclosing, world-building, and world-ordering concept of global responsibility and its supporting concepts alone are capable of keeping open the prospect of a global cosmopolitan and ecologically sustainable planetary sociocultural form of life.

Piet Strydom
University College Cork, Ireland

Preface

The words *justice* and *truth*, amid a world that habitually neglects these things and utterly derides these words, are nevertheless among the very greatest powers the world contains.

Charles S. Peirce

So if you want to know the truth about the universe, about the meaning of life, and about your own identity, the best place to start is by observing suffering and exploring what it is.

Yuvan Noah Harari (2018),
21 Lessons for the 21st Century

Responsibility is a foundation and a key challenge of our lives and in our interconnected world as it invites us to cultivate new ways of thinking and being. *The Calling of Global Responsibility: New Initiatives in Justice, Dialogues and Planetary Realizations* deals with this challenge. It strives to understand the challenge and calling of global responsibility as it emerges from new initiatives in global justice and dialogues.

The book builds upon a study I carried out on the subject of global responsibility from 2012 to 2016 which was supported by Indian Council of Social Science Research and I thank friends associated with it for their support. Mr. Biswaranjan Jena, Sugandha Jain, Dr. Athar Pirzada Hussain, Ms. Sujata Choudhury, and Ms. Julie Geredien have helped in this work formally as well as informally and I offer my thanks to them. I thank colleagues in our Institute for their support especially Professor Shashanka Bhide, the then Director of our Institute, and Mrs. T. Maheshwari—the then office secretary—in our Institute.

I dedicate this book to Philip Quarles van Ufford, Dipesh Chakraborty, Jill-Carr Harriss, Enrique Dussel, Pradjarta, Sailesh Rao, and Sudha Sreenivas Reddy. Philip Quarles van Ufford is a deep anthropologist of our times who has been concerned about moral critique and reconstruction of development. The book has its origin in the essay I originally prepared for our c-edited book, *A Moral Critique of Development: In Search of Global Responsibilities* (London: Routledge, 2003). I have learned from our journey together with Philip and he helps us think anew about our visions and practices of global responsibility. Dipesh Chakraborty is a deep thinker of our times whose many works including his latest, *The Climate of History in a Planetary Age*, help us in envisioning and embodying our layers of global and planetary responsibility. It has been enriching for me to have met with Dipesh first in his home in Chicago in 2003 and then have been in communication with him. Jill-Carr Harriss is an inspiring scholar-activist of our times who works on many frontiers of global solidarity. She works with Ekta Parishad, a Gandhian movement, and has been nurturing Jai Jagat walk and campaign. At present, she is creating a network for a non-violent economy in South India. It has been inspiring to walk and learn with Jill Behen and get new strengths and insights to continue to work for creating a better world for all of us. Enrique Dussel is a deep philosopher of our times whose work on the philosophy of liberation invites us to cultivate new ways of liberating our globe and creating a better world for all of us. It has been enriching for me to meet this great mind and kind soul once in Luxembourg in 2010 and to walk and meditate with his thoughts. Pradjarta is a silent and courageous scholar-activist of our times who founded Percik in Salatiga, Indonesia, as a civil society think tank of democracy and creative social action during the dark times of Soeharto's military dictatorship. The book discusses his work as well as that of Percik in democracy and inter-religious dialogues. It has been enriching for me to have met with Pradjarta and learn together with him as well as other friends of Percik such as Ambar, Agung, and Yani. Sailesh Rao is a silent and dedicated planetary yogi of our times who is working consistently for new pathways of climate yoga. We first met on a flight from Bhubaneswar to Delhi around 2010 and then in Los Angeles in 2018. The book discusses his work on climate yoga and his invitation for all of us to be butterflies rather than caterpillars. It has been enriching for me to have met with and learned with Sailesh Bhai. Sudha Sreenivas Reddy is a passionate and meditative social activist of our times who works on gender and communal violence and creates a new charter of human responsibility. It has been enriching for me to have met with Sudha more than two decades ago in Bangalore and then walk and learn together with her.

I am grateful to Professor Aakash Singh Rathore, Co-Editor of the Routledge Series, *Ethics, Human Rights and Global Political Thought,* for nurturing this book in this Series co-edited with Professor Sebastiano Maffletone. I am grateful to my dear and respected friend Piet Strydom for writing Foreword to this book and to my dear and respected friends Fred Dallmayr, P.V. Rajagopal, A. Osman Farah, Jeffrey Haynes, Julie Geredien, Sapir Handelman, Sabelo J. Ndlovu-Gatsheni, Mahmoud Masaeli, Beatriz Bassio, and Bian Li for joining us in the Epilogue roundtable conversations on this book. I am also grateful to Mr. Aakash Chakraborty at Routledge India for his kind support and encouragement. I thank Maanya Rao of New York University who worked with me as an intern for the summers of 2018 and 2020 for her research, writing, and editing help with this work. I also thank Swathi Suresh for her kind editorial help with this work. I hope that this book helps come to terms with our global challenges and embody new visions, practices, and movements of global responsibility.

Kartika Purnima and Guru Nanak Jayanti,
November 8, 2022
Ananta Kumar Giri
Chennai

1 The Calling of Global Responsibility

An Introduction and an Invitation to Adventures of Ideas and Transformative Movements

The Ganges of rights originates in the Himalayan of responsibilities.
—Mahatma Gandhi

Inheritance is never a given; it is always a task. It remains before us.
Jacques Derrida (1994), *Specters of Marx*, p. 54

The contemporary crises is a crisis of a particular *modelo civilizationio*, or civilizational model, that of Western capitalist modernity.
Arturo Escobar (2018), *Designs for the Pluriverse: Radical Interdependence, Autonomy and the Making of Worlds*

Globalization was on everybody's lips not so long, but the failure of globalization has led to nationalist upsurge in many parts of the world, such as the USA, Britain, and India. However, some crucial issues related to both valourized discourses of globalization and now anti-globalization such as justice, responsibility, and dialogue have received little attention in both theory and practice on the part of advocates and critics of globalization. It is by now incontrovertible that we live in a more globally interconnected but fractured, contentious, and fragmented world but what should be the character and direction of this evolving globality? Should globalization mean only economic globalization, and even corporate globalization, or should it mean humanization and planetary realizations? Corporate globalization with its slavish surrender to technology and profit maximization presents us a narrow view of human person in terms of economic self-interest and technical mastery. But globalization as humanization strives for a fuller human realization and for integral development (spiritual, political, economic, and social) of self and society. Globalization as humanization seeks to ensure universal self-realization of each of us. Universal self-realization refers to processes by which all of us can realize ourselves in terms of elementary practices of human development such as food and freedom and realize our highest potential. Planetary realizations refer to a new world realization that all of us are children of Mother Earth and it calls for transformation of existing boundaries, of rationality, nation-state, and anthropocentrism (Chakraborty

DOI: 10.4324/9780429347481-1

2021; Das 2008; Giri 2006b, 2018d; Kung 1996; Melucci 1996; Nussbaum 2006). Planetary realizations involve post-national transformations which go beyond the boundaries of nation-states and are reflected in contemporary transnational citizens' movements around the world for peace and justice. It also involves transformation of anthropocentrism suggesting that this world does not belong only to humans, and human beings are now challenged to overcome their human primacy and embody responsible guardianship of all species on earth (see Haraway 2016; Latour 2017, 2018; Chakraborty 2015, 2021).

Planetary realizations call for new initiatives in responsibility—global responsibility. According to Piet Strydom (2000, 2002, 2018), who is one of the few sociologists to have, reflected on the calling of responsibility for sociology and the wider public at large, the current challenges such as risks of environment, climate change, and new technologies call for responsibility as a new frame of engagement. For Strydom, "The theory of justice is today making way for another, still newer semantics form of the moral theory of responsibility which is crystallizing around a number of intertwined debates about the problem of risk" (Strydom 2000: 20). For Strydom, the rights frame had emerged in the early modern revolutions, for example, the Revolt of the Netherlands, the English Revolution, and the French Revolution. The justice frame had arisen in the wake of industrial revolution in "late eighteenth century England and continued unabated yet in a sublimated form until the second half of the twentieth century, focused on the problem complex of exploitation, pauperisation and loss of identity" (Strydom 2000: 20). These two discourses have inspired and influenced socio-political movements in the modern world, but now these two frames of rights and justice need to be transformationally supplemented by the vision and frame of responsibility. In his recent reflections on the age of responsibility and the prospect of a responsible society, Strydom argues that recent turn to responsibility has primarily meant the hegemony of "unencumbered individual responsibility" and we need to be part of social and cultural learning so that we can also embody collective responsibility (Strydom 2018). Strydom also argues that for building a responsible society, we need a "reflexive turn of responsibility" (ibid: 108).

This calls for work on both self and society. Modes of responsible engagement not only emerge from the public sphere but also involve appropriate self-formation and practices of self-cultivation, meditative reflections, self-transformations, and mutual transformations including spiritual mobilization of self and society.[1] Responsibility includes initiatives in socio-political as well as socio-spiritual responsibility. It includes both the ethical and the aesthetic and embodies what philosopher Karl-Otto Apel (2000) calls co-responsibility. For example, in the field of human development, state is not the only actor which is expected to be responsible. In my work on development ethics and social development, I have argued that the field of development consists of the work of actors such as state, market, social movements/

voluntary organizations, and self, and in this all the actors are called upon to embody a mode of responsibility (Quarles van Ufford and Giri 2003; Giri 2015; Taylor 2011).

The book builds upon such transformed understanding of our fundamental categories of thinking, acting, and understanding such as responsibility, as it hopes to contribute to new ways of practically and theoretically understanding these themes such as responsibility and justice (see Jung 1999). The book builds upon new works on justice such as global justice by Thomas Pogge (2001), Amartya Sen (2009), Sebastiano Maffletone & Aakash Singh Rathore (Maffletone and Rathore 2012), and Rainer Forst (2017) and justice for non-human animals and the disabled by Martha Nussbaum (2006). But in prevalent theories of justice such as John Rawls', there is a dualism between the political and the moral and in my study while carrying out studies on new initiatives in justice, I also rethink available theories of justice (Heller 1987; Rawls 1971; Sandel 1982). Along with overcoming the dualism between the political and the moral in realizing justice, there is also an epochal need to include both transformational self and social institution in thinking about justice (Sen 2008, 2009). Furthermore, in realizing global justice, we need to go beyond Rawlsian conception of "international justice" as "laws among peoples" (cf. Rawls 2001) as realization of global justice involves "person-to-person relations" not only "inter-societal relations" (Sen 2002: 48). As Amartya Sen challenges us:

> [J]ustice across borders must not be seen merely as "international justice." . . . A feminist activist in America who wants to help, say remedy some features of female disadvantage in Africa or Asia, draws on a sense of identity that goes well beyond the sympathies of one nation for the predicament of another.
>
> (ibid: 48–49)[2]

Sen argues that we can contribute to realizing global justice by developing "impartial spectator" in us as suggested by Adam Smith (Sen 2012: 138). Though Sen uses the word "impartial spectator" from Smith, he is pointing to a need for a more involved participation in the works of justice which is really that of concerned participants.[3] For realizing justice, we need to realize the complex relationship between impartial viewing and concerned participation, sharing and justice, and love and justice. This is suggested by seekers and philosophers such as Paul Ricouer (2000) and Fred Dallmayr (2007).[4] We need to bring works of love and works of justice together as we also need to bring together love and responsibility. Both of these require not only institutional and epistemic procedures in policy and politics but also all for work of faith and leap of faith in the sense Kierkegaard talks about it and in the way we see this manifest in societies and histories as in the works of Gandhi and Martin Luther King Jr. Justice and responsibility here becomes in a Kierkegardian sense concerns of truth and not just concerns of facts for which one becomes prepared to live and die (see Varughese 2012).[5]

Reflections on transnational justice are now illumined by engaging with the seminal works of critical philosopher Rainer Forst (2017) on this issue. Sen speaks about realization of global justice with and beyond institutional means such as Rawlsian nation-state and here it is helpful to bring Forst's perspective on transnational justice as non-domination into our conversation. Forster argues that we need to overcome domination in its multiple forms in our world, and for this we cannot just confine ourselves to the prevailing international system. We need to create new movements of moral and political constructivism so that we can problematize domination wherever it occurs and in doing so make concerned people critics and makers of their own laws of peoples. Forst differentiates his approach to transnational justice as non-domination from Philip Pettite's Republican theory of transnational justice as non-domination. For Forst, "In Pettit's republican theory, citizens are 'law checkers' interested in securing their freedom of choice, not 'law-makers' as in a Rousseian or Kantian sense" (Forst 2017: 161). Critics of global justice such as Neera Chandhoke (2012) point to the problems of paternalism in the reigning discourses of global justice where the recipients of global justice are never realized as autonomous and relational agents. Here Forst building upon both Marx and Kant argues:

> The dignity of a free person can never be understood merely in terms of the "enjoyment" of freedom or of certain liberties; it is always also a matter of the freedom of giving laws to oneself, the freedom of normative self-determination. This is a kind of freedom that comes in two modes—one moral and one political—but its *modus operandi* is the same, despite the difference between these two modes. The laws that constitute this practice and laws that are generated through it do not only protect freedom—they also express freedom.
>
> (Forst 2017: 157–158)

Forst further argues:

> Justice articulates the fundamental claim not to be determined but instead to be an agent and equal authority of justification that no one should be subjected to norms and social relations that cannot be justified in appropriate terms towards him or her.
>
> (Forst 2017: 130)

Forst's discourse theoretic approach to transnational justice opens up the field of global justice and helps us also to rethink power and justice. Forst offers a noumenal view of power—"to have and exercise power means to be able—in different degrees—to influence, use, determine, occupy, or even seal off the space of reasons for others" (Forst 2017: 42). But global justice involves not only social justice but also ecological justice (Clammer 2016).

Here we can link Forst's notion of "noumenal power" to Clammer's concept of ecological self which takes us beyond the notion of social self and social power to an ecological self arising out of deeper realization of one's being in nature which in turn helps us to transform power as control and domination to one as having the ability to work in concert with others, as Hannah Arendt (1958) suggests. Ecological self calls for identification with nature as well as suffering of both human beings and nature.[6] Transnational justice as non-domination also involves non-domination of Nature—both inner and outer—and it calls for development of ecological self along with other related movements such as planetary citizenship (Clammer 2016). Transnational justice also includes climate justice which involves rethinking global justice not only in anthropocentric terms but also in terms of justice for non-human species (see Nussbaum 2006; Haraway 2016).

Building upon Forst and relating it to our earlier conversations on this theme, realizing justice calls for working with and for justice in which state is not the only actor. All of us need to be strivers of justice and it involves justice works in our lives. Similarly, engagement with rights calls for rights works. But we need to link works to meditations, as a one-dimensional concept of action without meditation is not adequate to our task of realizing justice. Meditations help us go deeper in our inner lives and in the integrally linked spheres of intersubjective and trans-subjective relations and the public sphere. It resonates with what Hannah Arendt (1958) talked about *vita contempletiva* which is an unavoidable companion of *vita activa*. Meditation also brings the work of prayer and aspiration to our justice and rights works. In a similar way, we can think of responsibility in terms of responsibility works and responsibility meditations which would create new dimensions to ideas such as "co-responsibility" discussed by Apel (2000), "reflexive turn of responsibility" discussed by Strydom (2018), and my discussion of responsibilization. Meditation can also enrich Forst's discourse theoretic approach to transnational justice as non-domination which can also be realized as realization of non-violence or ahimsa (also see Patomaki 2019). This calls for developing non-violence in relations as well as non-injury in modes of thinking (Mohanty 2000).[7]

Meditation involves listening. Responsibility involves not only discursive argumentation but also listening (see Boyd 2017). It calls for listening to unheard voices and those who cannot speak. It also involves listening to higher self within oneself and others and to what Gandhi called the small voice within. Responsibility calls for development of our capacity to listen and competence to argue. In many projects of justice such as restorative justice, listening is emphasized. Listening develops our capacity to be attentive to others. In developing our capacity to listen, we become apostolic in our thoughts and actions as Pope Francis would tell us, and responsibility calls for such apostolic acts and thoughts of listening and attentiveness.[8]

For realizing responsibility, we need responsibility works and responsibility meditations. Responsibility is a multi-dimensional journey of realization and transformations and involves self, other, society, polity, and the world in manifold processes of co-realizations (Giri 2012). It includes both actions and meditations embodying what can be called meditative verbs of co-realization. It also involves initiatives in dignity, beauty, and dialogues. Dignity touches upon familiar themes, discourses, and movements in rights and justice, but dignity struggles go beyond discourses of rights and justice and point to the indispensable dimension of self. Dignity struggles involve both political and spiritual dimensions in a much more interconnected and integral way than movements of rights and justice in their conventional conceptualization and historical manifestation. So we need to conceptualize and realize responsibility as well as justice as meditative verbs of co-realizations of our realities and potential as well as emergence of new visions, movements, institutions, and practices.[9]

Responsibility as meditative verbs of co-realization as part of a dynamic movement of emergence includes compassion and confrontation. Compassion extends the discourse of rights and justice as we know and brings to it dynamics of generosity.[10] Responsibility, like justice and rights, involves epistemic works—ways of knowing, rather creative ways of knowing. Here to know is not only to know of but also to "know with," as R. Sundara Rajan, the deep philosopher from India, would tell us (Sundara Rajan 1998). In fact, knowing here becomes part of a process of what can be termed and realized as "knowing together with compassion and confrontation" (Giri 2013). Responsibility involves epistemological work, but this epistemological work is not bound within available epistemes; it needs to go beyond dominant epistemologies such as Western epistemologies which have constructed the world in a violent way of killing many different ways of knowing, being, and Earth-creating both epistemicide and Terricide (de Sousa Santos 2014; Escobar 2021). This journey of the epistemological beyond epistemology is also a journey of cognitive justice (Visvanathan 2017). Cognitive justice includes epistemic freedom, but it also challenges us to move towards cognitive and epistemic responsibility (see Ndlovu-Gatsheni 2018). Epistemology is not just epistemic in the modernist sense of being closed within its own procedures of certainty; it involves manifold processes work on self and self-transformation which can be broadly called ontological in an open sense. Responsibility involves simultaneous works and meditations of epistemology and ontology where both of them are simultaneously practical.[11] Responsibility as meditative verbs of co-realizations is part of a dynamic movement of what can be called ontological epistemology of participation (Giri 2006a, 2017). As this movement involves bringing together many related processes, dimensions, and actors, it is also related to aesthetics of border-crossing and establishing connections with and beyond disjunctions (Giri 2006a; Latour 2005).[12]

In a related way, responsibility involves both the subjective and objective. The objective constitutes the unavoidable challenge of our commonalty and commons which calls for a way of knowing and being with and beyond our subjectivity (see Frankl 1967). But this objectivity is not one of fixed objectivity or fixed positionality. Objectivity here emerges out of our trans-positional journey across positions and it is not a reproduction of positional objectivity as positioned objectivity (Sen 1993).[13] Much of rights and justice discourse quite rightly continue the struggle of positions, but there is also the responsibility of listening to and cultivating other positions while struggling for rights and justice. Reponsibilization involves transpositional movements, listening to others' positions and dances across positions (Giri 2016, 2023). But transpositional movements carry with them normative questions as well. While moving from positions to positions with empathy, it, at the same time, asks whether positions embody normative frames and modes of rights, justice, and responsibility or they are "instances of problematic justice" (Habermas 1990).

Here the works of some other contemporary thinkers on justice and responsibility deserve our careful consideration. In his important work, *Justice for Hedgehogs*, Ronald Dworkin (2013) tells us how we have to pay equal concern for the fate of each person and respect the responsibility and rights of each person. Dworkin links responsibility to interpretation. For Dworkin,

> We are morally responsible to the degree that our various concrete interpretations achieve an overall integrity so that each supports the other in a network of value that we embrace authentically. To the extent that we fail in that interpretive project—and it seems impossible wholly to succeed—we are not acting fully out of conviction, and so we are not fully responsible.
>
> (Dworkin 2013: 101)

Dworkin urges us to realize: "interpretation engages history, but history does not fix interpretation" (ibid: 300). In his review of Dworkin, Gerarld Droppett tells us that "Dworkin treats interpretation as a practice in the lives of individuals essential to the achievement of responsibility." Furthermore, Droppett tells us that

> Dworkin reads this interpretive problem as the apparent conflict between the two principles of human dignity—(1) self-respect, implying the objective importance of every one's life, and (2) authenticity, or each person taking individual ethical responsibility for her own life.

Dworkin urges us to realize the link between responsibility and interpretation but our interpretation needs to be non-biased, and for this we need to cultivate visions and practices of responsible interpretation. For this we need to move beyond interpretations of justice and responsibility from our fixed and initial positions and interpret these by moving to the positions of others. This corresponds to multi-topial hermeneutics discussed earlier. We also need to cultivate visions and pathways of interpretation which are simultaneously subjective and objective and embody transpositional subjectobjectivity.

Here works of political theorist Iris Marion Young are important. In her work *Responsibility for Justice*, Iris M. Young (2011) offers a social connection model of responsibility. For Young,

> The social connection model of responsibility says that individuals bear responsibility for structural injustice because they contribute by their actions to the processes that produce unjust outcomes. Our responsibility derives from belonging together with others I a system of interdependent processes of cooperation and competition through which we seek benefits and aim to realize projects. . . . All who dwell within the structures must take responsibility for remedying injustices they cause, though none is specifically liable for the harm in a legal sense.
>
> (Young 2011: 105)

Young speaks about political responsibility. For Young,

> political responsibility is not about heroism or charity. Rather, it requires a kind of solidarity that must be forged between individuals who may have little in common but for a preparedness to engage in a public debate and collective action for the sake of preventing structural injustices.
>
> (Young 2011: 350; see Demeterio *111* 2018)

Political responsibility has a spiritual dimension and it calls for new initiatives in education and dialogical spirituality. We get this invitation and insight from the works of Edith Stein, Martin Buber, Emmanuel Levinas, and Jacques Derrida, among others (see Baker 2018; Collins 2018). Edith Stein was a philosopher and educationist from Germany who challenged the rising Nazi power then. She emphasized upon cultivating empathy and emotion in education and cultivation of the whole person for leading a responsible life. As Baker writes:

> At the heart of Stein's philosophy lies a key to deciphering those elements of education which bring the desired fruit of responsible and wholesome persons in society. The heart is Stein's ultimate concern for discerning the truth of the human person in relationship, and during these most crucial years of her educational work, that truth she sought was inseparable from wisdom, Being, and desires of God.
>
> (Baker 2018: 163)

Martin Buber knew Stein. In his *I and Thou*, Buber tells us about the difference between I–Thou relationship and I–It relationship. For developing a life of responsibility, we have to develop an I–Thou relationship which involves our whole Being. For Buber, while the "primary word *I-Thou* can only be spoken with the whole being," the "primary word *I-It* can never be spoken with the whole being" (Buber 1958: 3). For Buber, I–Thou relationship is a field for realization and manifestation of Spirit: "Spirit in human manifestation is a response of man to his *Thou*. Spirit is not in the I, but between *I* and Thou" (ibid: 38). But we can also animate our I–It relationship also with a spiritual approach and realization which can help us in our much-needed journey of going beyond anthropocentrism. Here, as Maurice Friedman helps us realize: "It is not enough for man to use and possess things [I-It]. He has a great desire to enter into personal relationship with things and to imprint on them his relationship to them. . . [I-Thou]" (Friedman 1965: 54). *I–Thou* relationship may be limited to the double contingency of the self and the other, and following our previous discussion, it needs to be part of the triple and multiple contingencies as well as transcendence of self, other, and the world.

Here the works of Emmanuel Levinas and Jacques Derrida are also important for cultivating new visions and practices of responsibility. For Levinas, for embodying responsibility, we always have to give priority to the other, look up to the face of the other which is not in front of us but above us. But while paying attention to the other, the self also needs to prepare oneself and here we need to cultivate simultaneously self and other in a mode of responsibility going beyond one-sided primacy of either of them. Derrida invites us to think about how difficult it is to be responsible; at the same time, we can always strive for its creative cultivation, unfoldment, and manifestation in our thoughts and lives.

Responsibility not only involves dialogues but also involves a rethinking of the commonsensical understanding of dialogue involving two to polylogues, involving one, two, and many. Dialogue is linked to contingency but not confined to the double contingency of self and other but it touches the triple contingency of self, other, and the world (Strydom 2009). Dialogue involves a hermeneutics of putting one's feet not only in self but also in the other, not only in one culture but also in another culture. Dialogue does not happen only with the homeland of certainty and security of self and culture; it involves the pathos and joy of walking and meditating with other cultures. For the philosopher, theologian, and spiritual seeker, Raimon Panikkar (2008: 15), "dialogue is a process of prayerful mutual encounter." Building upon Panikkar's seminal work, Boaventuara de Sousa Santos tells us about pathways of *diatopical* hermeneutics:

> The aim of *diatopical* hermeneutics is to maximize the awareness of the reciprocal incompleteness of cultures by engaging in a dialogue, as it were, with one foot in one culture and the other in another—hence its *diatopical* character. *Diatopical* hermeneutics is an exercise in reciprocity among cultures that consists in transforming the premises of

argumentation in a given culture into intelligible and credible arguments in another.

(2014: 92)

Panikkar tells us further about *diatopical* hermeneutics:

> I call it *diatopical* hermeneutics because the distance to be overcome is not merely temporal, within one broad tradition, but the gap existing between two human *topoi*, "places" of understanding and self-understanding, between two—or more—cultures that have not developed their patterns of intelligibility. . . . Diatopical hermeneutics stands for the thematic consideration of understanding the other *without assuming that the other has the same basic self-understanding.* The ultimate human horizon, and not only differing contexts, is at stake here.
>
> *Diatopical* hermeneutics is a hermeneutic that goes beyond traditional *morphological* hermeneutics and diachronical hermeneutics, inasmuch as it "takes as its point of departure the awareness that the 'topoi', locations within distinct cultures, cannot be understood with the tools of understanding from only one tradition or culture" ("Autobiografia intellectual"). *Morphological* hermeneutics deciphers the treasures (*morphe*, forms, values) of a particular culture, a single tradition. *Diachronical* hermeneutics represents mediation between temporally distant eras in the cultural history of humanity, but still, normally, with reference to a single tradition.
>
> Seeking, among other things, to break out of the *hermeneutic circle* created by the limits of a single culture, *diatopical hermeneutics* attempts "to bring into contact radically different human horizons", traditions, or cultural locations (*topoi*) in order to achieve a true *dialogical dialogue* that bears in mind cultural differences. It is the art of arriving at understanding "by going through these different locations" (dia-topos). To achieve this, there must be a renewed encounter between *mythos* and *logos*, between subjectivity and objectivity, the heart and the mind, rational thought and the spirit that flies free breaking all rigid mental schemes.[14]

In the aforementioned paragraphs, Panikkar is challenging us to bring the morphological and topological—form and *topoi*—mythos and logos, subjectivity and objectivity, the heart and mind together going beyond our "rigid mental schemes" which poses challenges of transformations to conventional cognitive critiques as well as hermeneutics.[15]

Panikkar here tells us about putting one's feet in cultures which invites us to practice foot work in landscapes of self, others, and cultures as part of our visions and practices of dialogues (Giri 2012). But hermeneutics involved in this dialogical journey means not only reading of texts and cultures as texts but also foot-walking with texts and cultures as foot walks and foot works. It also means walking and meditating with cultures and texts as foot-working meditation while, as Thoreau (1947) would suggest, we walk like camels and ruminate while walking. Such foot walking, foot working, and foot-meditating dialogue transforms hermeneutics itself into a manifold act of democratic and spiritual transformation which involves related processes of root works, route walks, root meditations, route meditations, memory work, cultural work, and cultural meditations.[16]

Dialogue involves *diatopical* hermeneutics but this need not be confined to our feet only in two cultures; it needs to move beyond two cultures and embrace many cultures. Though physically we have two feet but we have multiple feet spiritually, mentally, and symbolically and dialogues call for us to walk with our multiple feet in many cultures and traditions giving rise to a *multi-topial* hermeneutics which involves not only creative foot work but also heart work (*herzwerk* as it is called in German). In multi-*topial* hermeneutics, we move from *topoi* to *topoi*, which helps us go beyond fixed positionality. It is also accompanied by moving from one time frame and temporality to other, for example, from the present to the past, from the tra-ditional to the modern, and from modern to the postmodern. Such a move-ment helps us go beyond the limitations of fixed temporality embody creative temporal plurality. It helps us overcome temporal blindness and appreciate the frames and world views of different epochs. Multi-temporal hermeneu-tics helps us in embodying responsibility to different times. This presents us a new hermeneutics of self, culture, society, and the world other than the conventional presentations presented to us such as that of conflicts between tradition and modernity or modernity and postmodernity (see Giri 2021c). *Multi-topial* and *multi-temporal* hermeneutics is animated by a multi-valued logic where different epochs and positions are not just opposed to each other in an either–or mode, but they also represent different dimensions of our inter-connected existence (see Giri 2002a, 2006a, 2016, 2018c, 2020a, 2021a & 2023). Multi-valued logic helps us in creative translation and communication across borders. Philosopher J.N. Mohanty (2000) tells us how multi-valued logic can build upon creative dialogues across philosophical traditions such as the Jaina tradition of *Anekantavada* which emphasizes many paths of Truth realization, Gandhian tradition of non-violence, and the Husserlian phenomenology of overlapping contents. In the pregnant thought of philoso-pher J.N. Mohanty which he crafts like a jewel:

> The ethic of non-injury applied to philosophical thinking requires that one does not reject outright the other point of view without first recog-nizing the element of truth in it; it is based on the belief that every point of view is partly true, partly false, and partly undecidable. A simple two-valued logic requiring that a proposition must either be true or false is thereby rejected, and what the Jaina philosopher proposes is a multi-valued logic. To this multi-valued logic, I add the Husserlian idea of overlapping contents. The different perspectives on a thing are not

mutually exclusive, but share some contents with each other. The different "worlds" have shared contents, contrary to the total relativism. *If you represent them by circles, they are intersecting circles, not incommensurable, [and it is this model of] intersecting circles which can get us out of relativism on the one hand and absolutism on the other.*

(Mohanty 2000: 24; emphases added)[17]

In the aforementioned paragraph, Mohanty tells us how we can build upon multiple philosophical and theoretical traditions of our world. Multi-*topial* hermeneutics involves movement across such multiple positions, locations, and traditions. It involves movements across positions which in turn involve empathetic as well as critical understanding of positions. In transpositional hermeneutics, we move across positions and interpret transpositionally and not just being imprisoned in our fixed and initial positions. Multi-*topial* hermeneutics is also confronted with the challenge of critique of our world views, cognitive frames, and power constellations as we move from one *topoi* to others. There is intractable question of power and knowledge in multiple *topoi* as we move from one *topoi* to others.[18] To come to terms with it, we need to bring together diatopical and multi-*topial* hermeneutics and critical inter-cultural hermeneutics together. In such a hermeneutic engagement, both the self and the other go through a process of critical self-displacement as well as self-distantiation. Critical philosopher Hans-Herbert Kögler (2007, 2014) here draws our attention to practices such as female genital mutilation in some cultures such as African cultures. Here a hermeneutic engagement with such cultural practices and horizons is meant not only to empathize with these but also to see the violence that is at work in these. Here one's entry as an outsider with one's critical epistemic and ontological perspective and way of living can help one to see the violence and annihilation that is at work in such practice which can also help the participants of such cultures especially the victims to see the play of power and violence in such practices and unfold their unrealized epistemic and ontological potential.[19] Critical self-distanciation in self leads to critical self-distanciation in other which leads to mutual determination in the direction of realization of beauty, dignity, and dialogues in self, culture, society and the world. In such critical multi-*topial* and transpositional hermeneutics, there is a change in positional subjectivity as well as positional objectivity of both the self and the other giving rise to transpositional suejctobjectivity which goes beyond the dualism between subjectivity and objectivity (Giri 2020b; Marotha 2009).[20]

A related engagement here is the vision and practice of reading and learning inter-religiously. Francis Clooney nurtures such a path of inter-religious learning. In his book *Learning Interreligiously: In the Text, in the World*, Clooney (2018) tells us how we can read texts, divine manifestations, and themes from different traditions together. For example, during the Advent period in Christianity waiting for celebrating the Advent of Lord Jesus Christ, Clooney tells us how we can read and experience the texts and processes of the birth of Krishna. Clooney here tells us:

Thinking about Krishna in Advent marks a way of practicing what we preach: interreligious learning is not merely a matter of ideas or confessions of faith aimed at one another, but it is a true intercultural exchange. By attentive study, we find our way into the literature of another religious tradition, we learn from it, and we consider in respectful detail what is said and how it is said. While this kind of study does not lead to answers to life's enduring questions, it changes us little by little, and we find ourselves to be Christians who have genuinely learned from another religious tradition. While it may not be possible for a Christian simply to believe in Krishna, for instance, there is no reason why a Christian, pondering the meaning of Christ's coming this Advent season, cannot learn greatly from how Hindus have interpreted the coming of Krishna into the world.

(Clooney 2018: 11)

For Clooney, "A willingness to listen to believers in other traditions and to learn from their theological reflection is also part of the great intercultural exchange to which we are invited in the twenty-first century" (ibid: 7).

We need to bring these concerns to our understanding of dialogue and responsibility. But this is also perennially confronted by intractable conflicts of interest, egoistic, and violent clinging to one's position and refusal to take part in dialogue leading to conflicts, disruption, destruction, agonies, and killing. So dialogue needs to acknowledge and find creative ways of overcoming intractable conflicts and propensity to violent disruption and destruction of dialogue. Dialogue needs to acknowledge the agonistic struggles of life and cultivate creative and transformational visions and practices of agonal dialogue and an agonistics of human existence—a dialogical agonistics of life.[21] Thus, multi-valued logic and multi-*topial* hermeneutics need to embrace immanent and transcendent normative questions embedded in our different locations and positions as well as our multi-*topial* and trans-positional movements within, across, and beyond these.

With these fundamental challenges of agonistics of human existence, it is essential to realize that dialogue, as polylogue, as it involves self, other, and the world, also involves a transformation of the idea of dialogue from mainly discursive to one that involves work and meditation, love and labour, and hatred and disjunction. In our conventional understanding, we look at dialogue mainly through participation in mutual discursive communication and participation in sharing of speech (Habermas 1990), but dialogue also is and can be non-discursive, trans-discursive, and meta-discursive as it involves love, labour, action, meditation, speech, silence, hatred, and disjunction. So we need to rethink and realize dialogue as meditative verbs of co-realization and pluralization which involve action and meditation, love and labour, and self, other, and the world in complex processes of interactions involving agreement and disagreement, violence and non-violence, and compassion and confrontation. About the non-discursive aspect of dialogue, we can here consider the example of movements such as Habitat for Humanity. Habitat is a movement from within American Christianity which builds houses with and for the low-income people in which people from different Christian denominations as well as non-Christians build together (see Giri 2002b). This act of building is an important dialogical act which may not involve discursive dialogue about religions and denominational beliefs. But nonetheless it involves dialogue albeit practical dialogues. This broad understanding of dialogue and the link between dialogue across borders, for example inter-religious dialogue and global responsibility, is made clear by Paul F. Knitter, a seminal thinker in this field. For Paul Knitter, "Religious traditions have the capability, if not the established record, of affirming global responsibility as ground and goal for inter-religious encounters" (Knitter 1995: 106). Knitter continues, "Different religions can and must share a global responsibility for eco-human well-being and justice" (ibid: 114).[22]

Inter-religious dialogue—dialogue across religious boundaries—faces the challenges of inter-cultural, cross-cultural, and transcultural communication such as prejudices and construction of the other as stereotypes, and it has to creatively overcome these challenges (see Akinade 2014).[23] Inter-religious dialogue as part of wider movements of dialogues across borders such as cultures, civilizations, and philosophies is an important part of global responsibility. But inter-religious dialogue not only involves interaction among religions, but also involves interactive processes which enable the participants to transcend their initial starting points and locations, thus becoming trans-religious (Giri 2020b). This study narrates the vision and experiences of several contemporary initiatives in dialogues such as inter-religious and trans-religious dialogues in India and Indonesia, multi-faith work in the USA, Parliament of the World's Religions as a space of inter-religious and cross-cultural dialogues, Dialogue of Humanity, Foundation of Universal Responsibility of His Holiness The Dalai Lama (FURHHDL), and charter for universal responsibility.

The processes of dialogues as well as realization of dignity involve not only ethics but also aesthetics. Aesthetics is the weaving thread in responsibility which helps us to undertake varieties of thread works amidst threat works. In our work on responsibility, beauty is in the middle as beauty, art, and aesthetics help us transform our consciousness and move beyond our fixed ideas of self and other. In thinking about responsibility, we thus need to bring ethics and aesthetics together in a spirit of transformations (cf. Ankersmit 1996; Ingold 2019; Quarles van Ufford and Giri 2003; Giri and Clammer 2017).[24]

Restorative Justice, Transitional Justice, and Responsibility

> *We brought the needle to sew the torn social fabric, not the knife to cut it.*
> Bantu proverb

One field of emergent justice where responsibility is manifested is restorative justice. Restorative justice also involves ethics and aesthetics in transformative ways and it is an important movement of transforming justice and a part of responsibility. Restorative justice tries to initiate dialogue between perpetrators of crime and victims. This happens both at interpersonal level and in wider inter-group levels.

Professor Chris Marshall teaches at University of Wellington, New Zealand, and he comes from a theological background. He has been a main scholar and activist of restorative justice in New Zealand which has also promulgated aspects of restorative justice in its legal system. I had met with Professor Marshall in May 2015 in his office in Wellington and he told me about his work in restorative justice. He told me how, in case of one accident victim, he created conditions of dialogues between victim's family and the perpetrator's family. In an essay of him, he writes, "While it contains retributive components, God's justice is fundamentally a restoring and renewing justice" (Marshall 2012). During our discussion, he agreed with me that work of restorative justice must be linked to creating training and capacity building on the part of all concerned individuals and institutions, such as police, court, civic groups, and common citizens.

Restorative justice also brings the issue of reparation and responsibility. In societies and histories, grave injustice has been perpetrated by forces of colonialism, slavery, and caste-based oppression. With regard to slavery, scholars have tried to work out a scheme of reparation. Professor William Darity Jr. of Duke University here offers an insightful scheme. He argues that, after the Civil War in the USA, Afro-Americans were promised land. But the then American President Andrew Jackson, who became President after the assassination of President Abraham Lincoln, reversed this. As a result, Blacks have suffered a great deal of material deprivation. They have gained some amount of civil rights, but they have been materially impoverished and marginalized. Darity proposes that in order to repair this historical injustice and contemporary deprivation, each Afro-American child should be presented a Government bond of say $50,000. This can be universal. So depending upon inherited wealth and income, this amount would vary. For example, a child born to Bill and Melinda Gates may get a bond of $50 while one born to an asset poor Afro-American may receive a bond of $50,000. This helps the child born to have good health care and education and also not carry the historical burden of suffering and deprivation. This scheme reflects the responsibility of society to bearers of historical injustice but it is not just confined to the victims. It is universal, providing a scope for unfoldment for everybody depending upon their present condition of deprivation and prosperity.[25]

Transitional justice is a related discourse and practice here that calls for our attention. Like restorative justice, transitional justice strives to create spaces and movements of justice and reconciliation and it is not confined only to demands of justice. Sang-Jin Han is an important scholar and creative actor for justice and reconciliation in Korea who has worked with former Korean President Kim-Dae Jong as well as former German President Richard von Weizsacker. In his introduction to the volume *Divided Nations and*

Transitional Justice: What Germany, Japan, and South Korea Can Teach the World, Han writes:

> The prime reason why we cannot be wholly satisfied with the justice-centered model is partly because it falls short of variability in some cases, but primarily because of the necessity of reconstituting a new political community open to perpetrators as well. From a legal perspective, justice may be an absolute value to be implemented. Politically, however, justice is a step towards reconciliation. Reconciliation is a higher value to be achieved through justice.
>
> (Han 2012: 12)

In his book, Han develops a communicative approach to reconciliation and here what Han writes discussing at length a new approach to realize reconciliation with both Japan and North Korea on the part of people of South Korea deserves our careful consideration:

> When we take the communicative approach to reconciliation, we are expected to see Japan not simply as a perpetrator who wronged us by waging war and military attack but also to draw attention to the Japanese way of thinking from their standing. For instance, we can pay sympathetic attention to the Japanese collective memory of war, asking why they see themselves as victims rather than offenders. We may then come up with the traumatic and catastrophic experience of massive suffering caused by the two atomic bombs dropped in Japan.
>
> Likewise, when we take the communicative approach to reconciliation with North Korea, we need to see North Korea not simply and narrowly from the perspective of systemic wrongdoings by the communist regime and understand why and how its citizens have seen their situations in the way they have. In this way, we can draw attention to the possible interaction between North and South Korea in the process of unification and transitional justice. The key issue is not about the legal, economic, administrative or military incorporation by one party or another but about the interactive process through which citizens of both sides come to share the common identity as well as the emotional and symbolic basis for a new political community to be constructed. . . . This communicative process is aimed at fostering mutual understanding and solidarity going beyond self-destructive consequences of national division.
>
> (Han 2012: 12–13)

This book has essays by Kim Dae-jung and Richard Von Weizsacker, former Presidents of South Korea and Germany, respectively. In his essay, "Power of Dialogue for Peace," Kim Dae-jung writes:

> Stretching back 37 years to 1971 when I ran for the Korean Presidency for the first time, I have been consistent in calling for peace in the Korean

peninsula: that Republic of Korea should engage in dialogue and cooperation with North Korea. With this position I vigorously opposed communism while advocating for peace and unification.

At that time, the world was swept in the Cold War. The voices of
hatred calls for the destruction of North Korea dominated South Korea.
Accordingly, my assertion prompted criticisms from the military regime
and its followers, and I was branded as a communist sympathizer. Consequently, I endured more than twenty years of persecution, including
imprisonment, exile, kidnapping even a death sentence.

Not daunted, however, I still advocated the "Sunshine Policy" based
on the three principles of peaceful coexistence, peaceful exchanges and
peaceful unification, which should be pursued consecutively in three
stages of a confederation, a federation and a complete unification.

When I came into office as the president of Korea in 1998, right from
my inaugural address I clearly set out this policy and proposed to Chairman Kim Jong-11 the holding of an inter-Korean summit to discuss
peace on the Korean peninsula and our own people's issues.

North Korea at first rejected my proposal as they thought I was trying to bring down the regime, in the same way the sunlight in the Aesop
fable succeeded in taking off the cloak of the traveler. But I persisted in
my efforts to persuade the regime, saying that "the Sunshine Policy" aims
are realizing peace and co-operation based on exchanges as well as bringing about a win-win outcome by means of dialogue between the South
and the North. Finally Pyongyang came to understand my intention and
Chairman Kim Jong-11 invited me to North Korea, where I had the
historic with Chairman Kim over three days from June 13 to 15, 2000.

(Kim 2012: 45–46)

President Kim Dae-jung narrates here his initiatives to realize peace
between North and South Korea. This is an aspect of responsibility across
borders. Such initiatives of responsibility or acts of responsibility create
beauty in societies and histories in place of ugliness. It creates beauty in lives
of people and becomes part of a broader social aesthetics. This challenges us
to understand beauty and aesthetics as an integral part of the movement of
and towards responsibility. Aesthetics constitutes a way of realizing responsibility as it becomes a part of responsibility itself as well as its very goal.
Aesthetics becomes a part of justice and rights as well. For example, in her
work *Poetic Justice*, Martha Nussbaum (1996) tells us how realization of
justice calls for not only legal approach but also poetic approach. Similarly,
we need to understand the poetic approach to rights where rights express
not only a hard core of a charter but also a poetry of aspiration, struggles,
and possibilities (Giri 2013; Visvanathan 1996). We can similarly talk of a
poetics of responsibility, aesthetics of global responsibility.

Here we can consider the following two examples as aspects of aesthetics
of global responsibility. *Reporter Sans Frontier* (RSF)—*Reporter Without*

Borders—is an international organization based in Paris and it reports about conditions of suffering and joy from around the world. During my visit to Paris in 2005 as part of fieldwork with the global justice movement of ATTAC, I saw an exhibition in a park in Paris organized by RSF which brought varieties of pictures from around the world. Many of these included suffering of people from Afghanistan. But the pictures were not only of misery but also of joy as people were striving to better their lives in the midst of varieties of challenges. These pictures create a field of perceiving the challenge of understanding global responsibility. These pictures represent reality but they are not only mimetic representations but also aesthetic representations which while creating a field of different perception and identification nonetheless challenges viewers for their own journey of self-development, mutual reflections, and participation (Ankersmit 1996). The noted philosopher and historian F.R. Ankersmit (1996) terms this aesthetic politics. We can also relate this to an aesthetic spirituality. Such aesthetic politics and spirituality becomes part of aesthetics of responsibility (see Wolf 2016).[26] Such an aesthetic politics and spirituality of responsibility is different from what David Harvey (1989) calls aesthetics of empowerment. In aesthetics of empowerment, one uses power to exercise domination over others and exploit and annihilate them and in the process derive sadistic aesthetic pleasure as it happened during Nazism and as it happens in many authoritarian regimes today. Aesthetic politics and spirituality help us overcome limits and perils of aesthetics of empowerment.

In this aesthetics of responsibility, art, music, and literature play an important role. During moments of struggle for rights and justice and during anti-colonial and anti-slavery movements, art, literature, and music had played an important role. We can remember here the crucial role of the novel *Uncle Tom's Cabin* in arousing people's consciousness about slavery. During Civil Rights movements, art, music, and songs had played an important role. Many of these songs such as the energizing and inspiring song *We Shall Overcome* emerged out of collective struggle of the participants.[27] Songs and music by singers such as Pete Seger and Joan Baiz also created a consciousness about suffering and the challenge of transformative participation across borders (Bartolf 2018). Novels and poems are also emerging now which point to the challenges of transformation of consciousness in the context of living in an interdependent world faced by threats of terrorism, climate change, and Corona virus (see Willis 2021).

Ten years after my being with the exhibition in Paris organized by Reporter Sans Frontier which created in me the thought about aesthetics of responsibility, on January 7, 2015, the office of the French satirical magazine Charlie Hebdo was attacked by two Muslim brothers in which the journalists and staff were killed. This killing was done to avenge the hurt created by publication of satirical cartoons of Prophet Muhammad (peace be upon him—PBUH from now onwards). Charlie Hebdo had earlier reproduced some of the satirical cartoons of Prophet Muhammad (PBUH) first published in

Danish newspaper Jyllands-Potsen on September 30, 2005. These cartoons back then had created a lot of suffering. Many people around the world had protested and around 250 were killed. But many people talked about it through the language of freedom of expression rather than through the language of responsibility. After the Charlie Hebdo killing, on January 11, 2015, about two million people, including Francois Hollande, the then President of France and more than 40 world leaders walked in a rally of unity and for freedom of expression. But none of the defenders and marchers uttered even a single line about responsibility of expression. They did not cry with many hundreds who were killed in protesting against these cartoons.[28] On the other hand, as the trial of alleged killers has recently begun in September 2020, French President Emmanuel Marcon emphasizes the need for defending the right to laugh and caricature:

> At the start of the trial of the attacks of January 2015, I say that to be French is to defend the right to laugh, jest, mock and caricature, of which Voltaire maintained that it is the source of all other rights.[29]

Recently there has also been heightened killing and violence on this issue. These cartoons were again reproduced by Charlie Hebdo with the starting of the trial of Charlie Hebdo killing in September 2020. This created anger and feeling of lack of respect and targeting on the part of Muslims in France and beyond. Added to this, Samuel Paty, a history teacher in a Parisian suburb, was beheaded by a Chechen assailant in front of the middle school where he taught after he showed Charlie Hebdo cartoons of Prophet Muhammad (PBUH) in a class on freedom of expression. After the attack, French President Emanuel Marcon said that this was "a typical Islamist terrorist attack," and that "our compatriot was killed for teaching children freedom of speech." But many countries such as Turkey and many individuals and groups around the world criticized this insistence of freedom of speech as insensible and offensive while condemning the murder of Paty. Days after the beheading of Paty, on October 27, 2020, three women in the church of Notre Dame in Nice were killed by a knife-weilding assailant. While in the neighbouring country of Belgium, an elementary school teacher has been suspended for showing a caricature of the Prophet Muhammad (PBUH) while discussing the beheading of Paty who had used the same image.

But here in asserting the right to freedom of expression and caricature as French values, as French President Marcon does, he seems not to realize the challenge of responsibility of expression.[30] Today in a globally interconnected world where in the name of being French can one one-sidedly justify mockery and caricaturing of the other without taking into consideration the pain and hurt it creates in both self and other? Should not one touch the hidden and unconscious dimension of the self which may still like to empathize with the positions and perspectives of the other? It seems that Marcon is neither able to understand the pain and suffering of others on this issue nor does he want to acknowledge this even after so much of blood has been spilt. Marcon seems unable to link rights to responsibility here which seems to be a failure of one-sided way of being French or being nationalistic or religious in this globally interconnected world of ours. Marcon asserts the so-called values of the French Republic but seems not to realize that Muslims living in France today are also citizens of France and they have as much right to constitute French Republican values today Marcon and others which would challenge us to realize not to use freedom of expression to disrespect others. Here theorists of Republicanism such as Forst and Pettit would challenge us to link Republican values to non-domination and respect.

Thus responsibility is a multi-dimensional challenge, and in this chapter, we have explored some of it. In the next chapter, we discuss about some movements of global justice, as they contain seeds of responsibility and transform justice.

Notes

1   Self-cultivation here includes self-limitation and self-transcendence which is also part of a process of mutual transcendence. In offering this linking thought I draw inspiration from Fred Dallmayr's interpretation of the thinking about self-cultivation in Gadamer. Dallmayr (2013) tells us that Gadamer interprets self-cultivation as not merely forming self within the given frame or convention of society but questioning and transcending this. In cultivating such a critical and interrogating path of self-cultivation, Gadamer draws inspiration from Hegel. As Dallmayr (2013: 63) tells us: "*Bildung* is not limited to the fine-tuning of existing capacities, but involves a movement of self-transgression in response to challenges." Such a critical path of self-cultivation can find a resonance with Habermas's discourse and pathways of post-conventional self-formation where self overcomes the limits of conventional mode of self-structuration and becomes post-conventional in its moral consciousness and communicative action. In his own creative ways, Piet Strydom also explores pathways of appropriate self-formation for our emerging world society which is helpful for cultivation of global responsibility at the levels of self, society, and institutions. For Strydom, "the subject is responsible for creating and constituting the sociocultural world" (Strydom 2020).

2   In their introduction to *Global Justice: Critical Perspectives*, Maffletone and Rathore tell us that Sen's person-centred approach to justice is a corrective to Rawlsian institutionalism or institution-centred approach (Maffletone and Rathore 2012: 9). In his own insightful essay on the book, "The Romance of Global Justice: Sen's Deparochialization and the Quandary of Dalit Marxism," Rathore (2012) argues that Sen's conception of global justice is rather romantic as it does not include issues of caste and class raised by Dalit Marxism in contemporary India.

3   Here what Sen writes deserves our careful consideration:

> While we are not likely to have a global state any time soon, Smith's emphasis on the use of impartial spectator has immediate implications for the role of global discussion in the contemporary world, to arrive at some comparative judgements. Global dialogue which I believe is central for world justice, comes today comes not only through institutions like the United Nations or the WTO, but much more broadly through the media, through

> political agitations, through the committed works of citizens' organizations
> and many NGOS, and through social work that draws not only on national
> identities but also other commonalities, like trade union movements, coop-
> erative operations or feminist activities.
>
> (Sen 2012: 138)

Sen in the aforementioned paragraph is pointing to important facts of many
movements dialoguing on global justice but Sen does not explore whether
these movements themselves take part in cross-cultural and cross-civilizational
dialogues on justice and peace. This book explores some of these issues.

4 Paul Ricouer argues that in justice we should not only be concerned with what
is our share of in a particular framework of distribution instead we should be
concerned with how do we share in. As he challenges us:

> The question is worth asking: what is it that makes society more than a sys-
> tem of distribution? Or better: What is it that makes distribution a means
> of cooperation? Here is where a more substantial element than pure proce-
> dural justice has to be taken into account, namely, something like a common
> good, consisting in shared values. We are then dealing with a communitarian
> dimension underlying the purely procedural dimension of the social structure.
> Perhaps we may even find in the metaphor of sharing the two aspects I am
> here trying to coordinate in terms of each other. In sharing there are shares,
> that is, these things that separate us. My share is not yours. But sharing is also
> what makes us share, that is, in the strong sense of the term, share in.
>
> I conclude then that the act of judging has as its horizon a fragile equi-
> librium of these two elements of sharing: that which separates my share or
> part from yours and that which, on the other hand, means that each of us
> shares in, takes part in society
>
> (Ricouer 2000: 132).

Ricouer also urges us to realize the link between love and justice which is also
emphasized in Dallmayr's work on justice. This integral link between love and
justice is also a helpful foundation for thinking about and realizing responsibility
as responsibility is founded not only facilitative law but also on love.

5 Kierkegaard makes a distinction between concerns of fact and concerns of truth.
We can realize responsibility as touching both fact and value in a border-crossing
manner where we are animated by concerns of truth. Tagore also presents a
related challenge of Truth realization here which is important to walk and meditate with:

> What you feel as the truth of a people, has its numberless contradictions,
> just as the single fact of the roundness of the earth is contradicted by the
> innumerable facts of its hills and hollows. Facts can easily be arranged and
> heaped up into loads of contradictions; yet men having faith in the reality of
> ideals hold firmly that the vision of truth does not depend upon its dimen-
> sion, but upon its vitality.
>
> (Tagore 1917: xii)

6 As Clammer tells us:

> The notion of cosmopolitanism makes little moral sense if it is divorced
> from the need for social justice that flows parallel to and constantly inter-
> sects with the need for ecological justice. What potentially unites them I sug-
> gest is the notion of the ecological self. If identification with nature is one
> half of that selfhood, identification with the suffering of that nature and of
> the humans and other entities that inhabit is surely the other.
>
> (Clammer 2016: 126)

7 Forst builds upon Marx and Kant in cultivating his theory of justice as non-domination but he can also build upon Gandhi and bringing Swaraj and non-violence to our theory and practice of justice (see Giri 2002a).

8 In Appalachia in the USA, I visited a project which is called Listening Project where the initiators had created processes where people can listen to each other. We also find such efforts in Peace and Reconciliation projects in different parts of the world.

9 Boaventura de Sousa Santos's work of sociology of emergence is helpful here. For Santos, "The sociology of emergences is the inquiry into the alternatives that are contained in the horizons of concrete possibilities" (de Sousa Santos 2014: 184). We can realize emergent alternatives as meditative verbs of co-realizations, not just nouns, and also as emerging from processes of meditative verbs of co-realizations involving different co-creators of transformations as well as the subjective and the objective, epistemic and ontological, political and spiritual. This process involves both compassion and confrontation. de Sousa Santos has drawn our attention to the significance of confrontation, especially creative confrontation, in giving birth to a different world. But we also need to cultivate compassion which has the courage to confront and confrontation which has integral compassion to self, other, and the world in its task of confrontation. We need to give birth to creative emergences as meditative verbs of co-realizations as works and meditations of compassionate confrontation. Responsibility and responsibilization emerge from such creative processes.

10 This can be related to Alasdair MacIntyre's project of "virtues of acknowledged dependence" (MacIntyre 1999). For MacIntyre, creative moral life calls for cultivating "virtues of acknowledged dependence" where virtues of justice and generosity come together. I have also been cultivating paths of bringing compassion and confrontation together involving compassionate confrontation and confrontational compassion (see Giri 2013).

11 Practical ontology and practical epistemology refer to the way both become practical through practices of love, labour, and meditation.

12 Herbert Reid and Betsy Taylor (2010) in their inspiring work, *Recovering the Commons: Democracy, Place, and Global Justice*, draw our attention to the significance of commons in rethinking rights and realizing justice which involves work of a new epistemology, ontology, and aesthetics. Building on John Dewey, they talk about the need for cultivation of an aesthetic ecology of public intelligence for realization of global justice and propose a folded ontology of individual in place of the flat of ontology of modern self what they call ecological ontology. This ecological ontology can help us realize rights as part of striving for common goods and ecology of well-being which can also be linked to what M.S. Swaminathan calls "ecology of hope" (cf. Swaminathan and Ikeda 2005). The spirit of ecology of hope as it can inspire movements of responsibility is explored in the following poem by the author:

Revolution: Greed, Grace and Transformation

Revolution!
Revolving around
Greed or Grace?
Mad and made for economic and technological revolution
Where is revolution of heart?
Turning of consciousness?
Can there be revolution
Without a movement?
Mass and energy
Body and spirit

Soul and society
For a new revolution of
Grace and Gravity
Compassion and Confrontation
Ecology of Hope
Moving together step by step
Towards a new science and spirituality of
Responsibilization and Transformation.
> [Written at Mata Amridananda Math, Vallikavu on
> the occasion of Amitavarsham 60, September 26,
> 2013, 9:30 AM. This is dedicated to Mata Amritanan-
> damayee and Dr. M.S. Swaminathan] (Giri 2019b: 5;
> also see Khairuddin et al. 2013)

13 Amartya Sen (1993: 130) talks about positional objectivity thus:

> Observations are unavoidably position-based, but scientific reasoning need
> not, of course, be based on observational information from one specific
> position only. There is need for what may be called "trans-positional"
> assessment—drawing on but going beyond different positional observa-
> tions. The constructed "view from nowhere" would then be based on syn-
> thesizing different views from distinct positions. The positional objectivity
> of the respective observations would still remain important but not in itself
> adequate. A trans-positional scrutiny would also demand some kind of
> coherence between different positional views.

14 This is retrieved from Raimundo Panikkar's website on the theme diatopical
hermeneutics on October 22, 2020.
15 This can open up further dialogues among the works and thoughts of Raimon
Panikkar, Piet Strydom, and Hans-Herbert Koegler.
16 I explore this in the following poem:

> Hermeneutics
> Is it only textual?
> Is it enough to turn to a hermeneutics of practice
> While making practice itself a text
> How does hermeneutics work with and beyond texts
> How does it become a foot work
> A meditation with soil, sole and sole
> Verbs of co-realizations
> Unfoldment of potential?
> > [Written on February 23, 2016, 5:30 AM.
> > Appearing in Ananta Kumar Giri (2022a: 31)]

The following poem by the author can also be read as hermeneutics involving
root work and memory work:

Roots and Routes: Memory Work and Meditation

Roots and Routes
Routes within Roots
Roots with Routes
Multiple Roots and Multiple Routes
Crisscrossing With Love
Care and *Karuna*
Crisscrossing and Cross-firing
Root work and Route Work

Footwork and Memory Work
Weaving threads
Amidst threats
Dancing in front of terror
Dancing with terrorists
Meditating with threat
Meditating with threads
Meditating with Roots and Routes
Root Meditation
Route Meditation
Memory Work as Meditating with Earth
Dancing with Soul, Cultures and Cosmos.
[UNPAR Guest House, Bandung February 13,
2015, 9:00 AM]

This is from Ananta Kumar Giri (2019b), *Weaving New Hats: Our Half Birth Days.*

17 Jaina tradition refers to *Anekantavada*, multiple perspectives of Truth. Building on this, I talk about *Anekantapatha*, multiple paths of Truth engagement and Truth realizations. Gandhi himself in his own understanding of this Jaina path calls his path practical. Here Margaret Chatterjee in her study of Gandhi tells us how in 1924 "Gandhi made a statement saying he had no objection to being called an anekantavadi or syadvadi, that it is to say, one who accepts many sidedness of truth." Gandhi adds here:

> But my *syadvada* is not the *syadvada* of the learned, it is peculiarly my own. I cannot engage in a debate with them. It has been my experience that I am always true from my point of view, and I am often wrong from the point of view of the honest critic. And I know that we are both right from our respective point of view. And this knowledge saves me from attributing to my opponents and critics. . . . It is this doctrine that has taught me to judge a Mussalman from his standpoint and a Christian from his. Formerly I used to resent the ignorance of my opponents. Today I can love them because I am gifted with the eye to see myself as others see me and vice versa. My *anekantavada* is the result of the twin doctrines of *satya* and *ahimsa*.
>
> (Gandhi quoted in Chatterjee 2005: 306)

18 Here what Vince Marotha writes deserves our careful consideration:

> In cross-cultural encounters the anticipation of meaning leads to boundary constructing processes in which the self and the other are equally involved in the construction of meaning. However, when cross-cultural encounters are unequal and based upon a power relationship then the anticipation of meaning on the part of dominant self becomes the standard with which meaning and understanding are constructed.
>
> (2009: 280)

19 Here what Kögler (2014: 277) writes deserves our consideration:

> The core idea of such a "critical hermeneutics of self-displacement" can be expressed in conceptual contrast to the principle of rational assimilation that I sketched above. If it were possible to understand differently situated agents without assimilating their views to ours, we could then employ the hermeneutic understanding of such views to look at ourselves from their perspective and thereby achieve a relative outsider position vis-á-vis our own taken-for-granted interpretive schemes. Inasmuch as many of our

usual beliefs and assumptions are intertwined with power relations, such a hermeneutics would, by displacing our interpretive schemes, contribute to a reflexive self-distanciation from our power-shaped social agency. This hermeneutic project would thus allow for the reflexive reassessment of our situation in order to achieve a higher degree of self-empowered agency and autonomy. Also, interpretive practices in the human and social sciences would support the normative project of reflexive self-determination.

20 Here Vince Marotha suggests that in hermeneutic encounters outsiders and strangers "dialectically adopt a frame of mind characterized as a 'subjective objectivity' which involves both remote and near, detached and involved, indifferent and concerned" (Marotha 2009: 277). Marotha here refers to the work of sociologist Robert Park and his conception of a hybrid self-emerging from cross-cultural encounter which involves a more complex and meaningful mode of understanding. Marotha also discusses the works of George Simmel, Zygmunt Bauman, and Peter Pels (a contemporary anthropologist in The Netherlands) where "social epistemology of stranger" gives rise to a "third position" (ibid: 278). We can relate to such a "third position" to transpositional subjectobjectivity of participants in hermeneutic meetings and encounters.

21 This can be related to the perspective of agonal democracy cultivated by Ernesto Laclau and Chantal Mouffe (1985).

22 Knitter here presents us the following transformational challenges:

> The challenges that face us in our suffering brothers and sisters and in our suffering planet are calls to religious persons to be not only *prophets* transforming the system but also *mystics* plumbing the depths of the Divine or the Real. To feel global responsibility, to give oneself to the task of struggling for *soteria* in this tormented world . . . to feel claimed by the sacredness of the earth and call to protect the earth—such human experiences and activities constitute a universally available locus, an arena open to all, where persons of different religious backgrounds can feel the presence and empowerment of that for which religious language seems appropriate. . . . Working together for justice becomes, or can become, a *communication in Sacra*—a communication in the Sacred—available to us beyond our churches and temples.
>
> (1999: 114, 115)

23 Piet Strydom (personal communication) here refers to the works of Henri Tajfel, William Doise, Jacques Demogon, Jan-Rene Ladmiral, Pascal Dibie, and Christian Wulf whose works on intercultural communication can be fruitfully studied together for a deeper understanding of inter-religious dialogues.

24 Piet Strydom, one of the few sociologists bringing to sociological discourse the significance of responsibility, also emphasizes the significance of aesthetics for cultivating responsibility in our lives and institutions (personal communication).

25 Professor Darity offered this in his lecture during Ambedkar's birth anniversary in Jawaharlal Nehru University, New Delhi, on April 14, 2016.

26 Philosopher Susan Wolf (2016) makes a distinction between aesthetic responsibility and moral responsibility. Moral responsibility is more rule-governed while aesthetic responsibility is more nuanced.

27 Guy and Candy Cavens are creative musicians and they are based in Highlanders, a centre for mutual dialogue and social transformations in Tennessee. In Highlanders, leaders of different movements of social change come together and seat together around a table and share their experiences. This experiential sharing helps them to learn together and get strength. Both of them were active participants in the civil rights movements. I met them during my visit to

Highlanders. Guy was telling me how the song *We Shall Overcome* emerged out of struggle. For example, the line "we are not afraid" in this song came up from the mouth of people who were taking part in a meeting in the Highlanders. Similarly, new songs and poetry come up during movements of global justice such as ATTAC and Ekta Parishad.

28 The following poem by the author explores this challenge of responsibility which may be read as an aspect of a poetics of global responsibility.

Weaving New Hats, Heads and Hearts of Love

My Green Cap
A Canopy of Care
Gone into Air
Lost for Ever
The Head of My Daughter is Gone
In the Dance of Battle
Our Sons are Gone
In Suicide Bombing
Our Fathers are Gone
In the Name of Expression
Freedom of Expression
Freedom of Expression
Tearing each other apart
Can we weep together?
Where is mutuality of expression?
Respect, Responsibility and Compassion?
How do we touch our humiliation?
You draw cartoons
The head of my Mother is gone in the wind
Monsieur President!
Oh President!
Oh Editors, Soldiers and Legislators
Do you cry for us?
Where is your tear?
Oh *Jihadis!*
Killing in the name of *Payagmbar*
The Prophet of Peace
Whoever stitches
Canopies of Respect and Co-Realizations
Hats, Heads and Hearts of Love
In the Deserts of Hatred and Retribution.
(Written during Jaipur Literature Festival,
January 23, 2015, 9:15 AM)
(see Giri 2019b)

29 The internet source about this puts it this way:

More than a dozen defendants went on trial this week for their role in the killing spree in the Paris offices of Charlie Hebdo, where 12 people died. This week, the magazine republished the caricatures of the Prophet Mohammad that had sparked the ire of Islamist militants.

(NDTV News, September 4, 2020)

30 Here what Judith Butler writes deserves our careful consideration: "[. . .] free speech should not take priority over rights to equality" (Butler 2021: xv).

2 Rethinking and Transforming Global Justice

ATTAC, Ekta Parishad, and Other Initiatives in Global Justice and Responsibility

Seattle, Prague, Nice, Gothenburg, Genoa, and Cancun—a few years ago these names would have been an arbitrary enumeration of towns. Today these places are widely known as synonymous for the global protest against neo-liberal globalization. Moreover, these protests mark the emergence of a new transnational social movement, which represents one the rather rare cases of transnational collective action.... Erroneously many journalists and some academics call it the "anti-globalization movement." . . . I prefer the term *global justice movement*, because it is more accurate. The global justice movement indeed opposes the current form of neo-liberal globalization, but it also supports global political solutions—such as the Tobin tax, designed to fight negative consequences of neo-liberal globalization and achieve social and economic justice on the national as well as international level. The origins of the global justice movement can be traced back to the events in the early 1990s, such as the uprising of the Zapatistas in Chiapas, Mexico or the mass protests against the 1988 International Monetary Fund (IMF) and World Bank meeting in Berlin.

Felix Kolb (2005), "The Impact of Transnational Protest on Social Movement Organization: Mass Media and the Making of ATTAC Germany," pp. 95–96

Having emerged a few years after the fall of the Berlin Wall, alter-globalization may be considered as the first truly "global movement," both because it gathered grassroots citizens from the north and from the global south, and because it has raised some of the major challenges of the coming global age. Along with INGOs, committed intellectuals and civil society organizations that had already become increasingly international . . . grassroots actors from the north and from the global south have played a key role in this movement since its early beginning. This has particularly been the case with small farmers and indigenous movements which have been key actors in several of the founding events, including the Zapatista uprising and the Asian farmers' mobilizations against the GATT. Their forms of activism and their poetical discourses have become a major source of inspiration for activists in the north and in the south. Their global networks are unrivalled, with Via Campesina uniting over a million farmers and becoming a central actor in global protests and Social Forums. With the issue of climate justice, both small farmers and indigenous movements are once again at the forefront

DOI: 10.4324/9780429347481-2

of the global struggle, not only because they are particularly threatened by global warming, but also because they have developed some of the most inspiring alternatives regarding this issue. Previously considered as anachronistic left-overs of a pre-modern era that would eventually disappear with the modernization process, small farmers and indigenous peoples are now at the forefront of a global debate around rethinking our development model, in a world that is less characterized by its perpetual expansion and growth than by the limitedness of the earth and of its natural resources.

—Geoffrey Pleyers (2010), *Alter Globalization: Becoming Actors in a Global Age*, pp. 261–263

Responsibility builds upon justice but, at the same time, is not limited to it. As we have discussed in the previous chapter, responsibility builds upon transformation of justice. Responsibility is a perennial movement of transformation, and thus it is better to understand and appreciate it as a process of responsibilization. Responsibilization as a process of self and social transformation includes both action and meditation. As we have discussed in the previous chapter, it is part of meditative verbs of co-realizations.

In this chapter, we discuss some movements of global justice which contain germs of global responsibility. We discuss ATTAC (Association for the Taxation of Financial Transactions for the Aid of Citizens) which is a movement for appropriate tax justice and the inter-linked broader movement of solidarity and some other movements such as Indignados and Ekta Parishad.

Movements of Global Justice in the Direction of Global Responsibility

ATTAC (Association for the Taxation of Financial Transactions for the Aid of Citizens)

ATTAC (Association for the Taxation of Financial Transactions for the Aid of Citizens) is one of the transnational social movements, which is struggling for embodiment of global responsibility in concrete ways. It demands that multi-national companies should pay a certain amount of tax to the local communities where they are working what is now called the Tobin Tax. To reduce volatility of financial transfer and consequent danger of financial instability to the host countries especially the weak ones, ATTAC demands that each such financial transaction should be taxed. Nobel Prize winning economist James Tobin had proposed this idea, and drawing upon this, ATTAC was fighting for administration of Tobin Tax. Here what political scientist Heikki Patomaki writes about Tobin Tax deserves our careful consideration:

Tobin Tax seems to promise a new phase in the politics of globalization. The Tobin tax is a low-rate tax on all currency transactions, which form an essential component of most activities in the global financial markets. A currency transanction tax would defend and develop the autonomy of states, and, in particular give more room for manouevre of their economic policies. However, whereas Tobin tax regime can be seen as defending some aspects of their state sovereignty, it also opens up new, path-breaking global ethico-political problems of governance. . . . [In the context of financial speculation] The best way of for most states in most contexts—and for many other political actors well—to regain control over these forces is to *organize actions globally*.

(Patomaki 2000: 78–79, 83)

Patomaki, who himself is an activist of ATTAC and the leader of ATTAC in Finland as well as a public intellectual in Finland, Europe, and the world especially in his participation in World Social Forum (more on it later), engages with the Tobin Tax as a leverage for democratizing globalization and global institutions. Patomaki (2001) urges us to realize the limits and possibilities of Tobin Tax. For Patomaki, while "the Tobin tax addresses the power of speculative financial markets and their socio-economic consequences," it fails to "address wider issues of economic *efficiency*, *justice*, *democracy* and human *emancipation*" (Patomaki 2001). For Patomaki, while James Tobin's original case was based on the ideals of *autonomy* of states and economic *stability*, now we need to make this part of a global frame of interdependence and governance. While Tobin and his followers initially wanted International Monetary Fund to collect the taxes of currency transactions, Patomaki argues that we need to create a new global institution called Tobin Tax Organization (TTO) which interested States can create as a global body and the membership should be extended to other interested actors such as civil society organizations (2001). Formation of such a global organization can be used to democratize globalization including financial globalization and it can be instance of creation of transnational justice as non-domination as suggested by Rainer Forst and discussed in our previous chapter in which interested actors take part in creating laws which are mutually binding as rule-making subjects. As Patomaki writes:

The beauty of the Tobin tax lays in its potential to give rise to new political constellations. It means both the autonomy of states and new global institutional arrangements aiming at, and leading to, democratising globalisation. In the irreversible historical processes of structuration, this new phase of globalisation would lead to new kinds of political sagas, too.

(Patomaki 2001)

Along with striving for realization of Tobin Tax, ATTAC was also fighting for Third World Debt reduction and against privatization of public utilities such as water. Since 2001, I have been carrying out conversation with

Figure 2.1 The symbol of ATTAC with its % sign.[1]

different local chapters of ATTAC around the world. Here I present my research with ATTAC over the years, as it pertains to themes of global justice and responsibility.

ATTAC started in a rather unexpected way. Bernard Cassen, a journalist with the French newspaper *La Monde*, had written an article in this paper discussing the problem of global financial flow, globalization, and the need for taxation. After reading this article, many people called him if they could join his initiative thinking that such an initiative already exists. This inspired Cassen and his friends to start a movement for this which came to be known as ATTAC.

I had a meeting with Cassen in his office in Paris in July 2005. During our conversations, Cassen said that he and his fellow strivers started with Tobin Tax, because it was an important point of entry. Tobin Tax would be a source of financing development. "We fight against all forms of neo-liberal globalization: Tax Heavens, WTO & GMO." Cassen further said:

> We define ourselves as an action-oriented popular education movement. Our aim is to uproot liberalism [neo-liberalism] from our mind. We have been infected by the liberal virus. It is not finance that should be in the commanding position. We are here to change people's mind.

ATTAC brought its vision of a different kind of Europe and the World to bear during the French referendum on the Treaty establishing a Constitution of Europe which was held on May 29, 2005. As the proposed Constitution promoted a neo-liberal Europe dominated by finance capital, ATTAC opposed it on the ground of its neo-liberation implication for economy and society. Before taking this oppositional stance, ATTAC organized an internal referendum and found that 84% of its supporters said no to the proposed Constitution of Europe. Thus, ATTAC opposed the constitutional referendum and played a crucial role in its defeat. During our conversation, Cassen clarified that he and ATTAC were not against EU, but they were against the neo-liberal takeover of Europe. They are for Solidarity Europe, Europe as a cooperation circle.

This vision of Europe as a co-operation circle can remind one of Gandhi's idea of society and the world as concentric circles with a different basis of organization of self, society, and polity.[2] Cassen may not have spoken about Gandhi during our conversation but many of the activists of ATTAC and many related global justice movements draw inspiration from Gandhi. Cassen also wanted to organize ATTAC as varieties of co-operation circles rather than one centralized structure. During our conversation, Cassen said that in France ATTAC consists of autonomous local committees and then a National Forum. It follows democratic path, but it is not just one of simple democracy of one member, one vote. The founding members of ATTAC constitute the Board of ATTAC and this is meant to prevent take over by new members. The issue of representation, democratic representation, is a complex issue not only in mainstream politics but also in social movements and movements for global justice.[3] Cassen deals with this challenge with a mixture of openness and privileging of the pioneering few. Different ATTAC chapters around the world deal with this differently as also other related movements such as Ekta Parishad. In his study of ATTAC and other initiatives in the Alter-Globalization movement, Geoffrey Pleyers (2010) tells us how many grassroots activists in such movements feel that there is democratic deficit in their organizations and less participation and representation in their organizations and dominance of a select few who present themselves as founding experts and saviours. For Pleyers, "Many small alter-globalization think tanks and NGOs are built by and around a committed intellectual who presides as lifelong president. Even larger organizations headed by intellectuals do not always manage to address adequately the question of internal democracy" (Pleyers 2010: 146). In his study, Pleyers tells us how there was an electoral fraud in the election of ATTAC France in which the faction led by founder Bernard Cassen was involved which shocked many. Pleyers writes about this:

> While it disappointed thousands of activists and surprised journalists, because ATTAC defined itself as a civil society organization which fights for a "reinforcement of democracy," this electoral fraud was inscribed within an extension of the logic which had dominated ATTAC-France since its inception and which is encountered in many organizations of

the way of reason: the resurgence of an avant-garde attitude by leading intellectuals. ATTAC-France was shaped by a top-down and authoritative approach to organizing a movement, in which intellectual leaders indicate the way forward without much consideration of internal democracy or local chapters. The distance between "intellectual" leaders of the organization and "grassroots activists" is taken for granted among the members of the executive team, as the following extract from an interview in 2001 with an ATTAC-France administrator, chosen by Bernard Cassen, indicates: There is obviously a difference between grassroots activists, who join ATTAC to get a political culture, and the members of the scientific council who are academics and directors of newspapers and magazines.

(Pleyers 2010: 146)

At the same time, Karl-Julius Reubke who has studied Ekta Parishad and also reflects upon the challenge of democracy in a movement like Ekta Parishad tells us how Ekta Parishad also follows a complex process of openness and the needed closure to make decisions and ensure stability and accountability of the movement itself (Reubke 2020).

During my visit to France in 2005, I had spoken with some other members of ATTAC in France. Christopher Ventuara is an important leader of ATTAC in France. Ventuara tells us that ATTAC France had 30,000 members and it had alliance with 1,000 organizations. In his words, "We are able to have a big echo in society and our ideas touch people. We are able to have a big echo in society." Though ATTAC collaborates with left-leaning political movements, ATTAC maintains its autonomy. As Ventuara says, "Our work is to create an influence on everybody, on public opinion. We do not want to be identified as a think tank for the leftist." Ventuara tells us, "We have our scientific council in ATTAC France. It consists of 150 members and includes sociologists, economists, philosophers and lawyers." ATTAC said no to the Constitutional referendum, because it did not want to be part of a neo-liberal remaking of Europe like the USA. Tells Ventuara: "What is at stake is a project of society, civilization. Do we build solidarity between generations?" "We are working as Europeans for another Europe." Regarding Tobin Tax he says:

We need to develop fiscal policies. We need to finance social needs. We need fiscal justice. We have to rethink tax system—global tax system, harmonization at the continental level. . . . Also in the North we are poor people. The money made by speculation at the global level is indecent. Speculation creates social damage. We need clear social and political control. Tax system can be a tool for social justice and not just bad thing.

I had a discussion with Ventuara about ATTAC's position on subsidies for French farmers which are having a negative impact on poor farmers of the world. Ventuara says, "Subsidies cannot be separated from international rules. Because we believe in food sovereignty, we have to protect our farmers. In the system, we cannot accept that our agriculture can be destroyed. The problem is international agriculture." At the same time, Ventuara says that priority should be for local production and consumption and not for global trade on agricultural commodities.

Agnes Perin is a member of the National Board of ATTAC in France. I had met with her on July 18, 2015. During our conversations, Perin told, "I joined in the winter of 1998–99 after attending some lectures about globalization. . . . I was part of a regional member of a local group. I became President of this group. I became more and more activist. I presented my candidature to be a member of the National Board. I was not elected. One year after, six members were nominated and I was one of them," Perin tells us how ATTAC plays an important role in education of the public about dangers of neo-liberal project. About her own local group, she says:

> We have 15 activists. We make work camps which are our summer universities in which we use participatory method, a mixture of culture and politics. We make political theatre on issues such as Mundial Bank/IMF, GATTS. Fee in the summer university is dependent on what you earn and is not for upper classes only. It can begin with as low as eight euro per participant.

About her own involvement in ATTAC, Perin says:

> I have a 35 hour work week but I spend a lot of time with ATTAC. It takes a large part of my hobby. But I do not think joining ATTAC has changed me. I was an activist before but joining ATTAC has helped to find ways to think about political topics.

Furthermore,

> I feel good that it is not a political party. No boss-no chief. You can think different no obligation to have the same position as others. . . . There is no obligation to think one-way. There is an independence of each part of ATTAC. There is no hierarchy. There is conflict but no war. The statute of ATTAC was for a little group but now we have 30,000 members. There is conflict between Founder College and Activist College.

Perin has a special love for literature and culture and wants a more creative relationship between culture and politics. She says, "Culture was not sufficiently important in ATTAC in the beginning. I have been trying to bring the cultural dimension to ATTAC's work by bringing theatre."

Perin says that ATTAC France, like ATTAC Germany, works closely with trade unions, but it is not subservient to them. It also does not take part in elections.

In recent years, ATTAC France has lost a lot of its grassroots members. It has also declined. I was speaking with Professor Alain Touraine, the noted sociologist of France and the world in his office at MSH (Maison des Sciences de l'Homme), Paris, in January 2013. He says:

> ATTAC has lost part of its visibility because of internal contradiction. Its strength came from grass-roots movement. It is difficult to translate economic critique to political movement. It is difficult to transform grass-roots movements to political action. Many grass-roots movements do not want to be subjected to political action. In the Chiapas the Zapatistas[4] did not want to be subjected to political language.

I had a meeting with an important leader of ATTAC in Paris in January 2013. He acknowledged that ATTAC has lost its earlier mass base. It has now become much more a centre of critical education. It publishes books on important aspects of the global economy.

ATTAC in Other European Countries and Around the World

ATTAC in Sweden

Flowing with the spirit of enthusiasm of the newly established ATTAC in France, ATTAC Sweden was established on January 6, 2001. It caught the imagination of many students and youth and created the possibility of a new radical politics. ATTAC chapters opened in major cities in Sweden, such as Stockholm, Gothenburg, and Malmo and students in local universities such as Gothenburg University played a major role in the spread of ATTAC in Sweden. In the beginning, one could witness a lot of enthusiasm and idealism. ATTAC played an important role in organizing non-violent protest against the EU summit held in Gothenburg in 2001. As part of this participation and resistance, Hans Abrahamsson, a teacher at Gothenburg University and a leader of ATTAC, here formulated and coined a new strategy of resistance what he called confrontative dialogue. Abrahamsson reasoned with his fellow resisters to realize that usually confrontation involves violence and lack of any communication between contending parties while dialogue in the conventional sense involves lack of serious disagreement and acceptance of a consensus. ATTAC should go beyond either of these options and practice confrontation which is non-violent and dialogue which involves contestations together.[5] As a way of nurturing confrontative dialogue, ATTAC Sweden organized several dialogues with leading ministers and political leaders of Sweden on important local and global issues such as debt reduction. For example, it organized TV discussion with the finance minister of the country on the issue of Third World Debt reduction in which the participant from ATTAC challenged him to lay bare his position on this topic before the public. With such an approach of confrontative dialogue, it was possible to listen to the position of the Minister and the Government on this issue and then to move beyond. One of the leaders of ATTAC Sweden from Gothenburg, Helena Tagesson, had a television debate with Goran Persson, then Prime Minister of Sweden.

With such a vision and practice of confrontative dialogue, ATTAC strove to lead the protest at the G8 summit in Gothenburg held in June 2001 in a non-violent way. It had organized several meetings between the local police officers and protesters of different hues including the ones which believed in violent protest and worked to put different protesters in different camps in coordination with the police. But with all these, there was violence in Gothenburg and police also unleashed a brutal attack on the protesters. There was a party organized *by Reclaiming Our Streets* and three of the protesters were injured though gunshots by police—one severely injured. Luckily, he survived. The media showed the images of violence in Gothenburg and people associated this with ATTAC though ATTAC had done its best to prevent it and violence was never ATTAC's philosophy or strategy. After this, membership of ATTAC Sweden declined. People continued to associate ATTAC with violence, and media portrayal remained continuously negative. This led to decline of membership which many of the leaders took it with a great deal of courage and stride. For example, Helena Tagesson, an important leader of ATTAC, with whom I was discussing this, took this as a challenge of creative organizational building and working on people's awareness as those who had joined ATTAC out of initial enthusiasm had left. But still the decline continued and now ATTAC, as in France, is a much-reduced actor and organization in the global justice movement locally and globally.

During my field work with ATTAC in Sweden since 2001, I had met with and spoken to many volunteers of ATTAC. Malinn was one such. She was a volunteer of ATTAC in Malmo as she was studying at the local Malmo University. She once took a bus load of fellow volunteers to a European Social Forum. She commented: "The bus itself became a social forum." Marcus Nilsson was another activist of ATTAC based in Malmo. He worked as a school teacher then. For him, ATTAC provided an alternative space for conversation and for politics. He says:

> Young people think that if you are part of an organization you become a small person. But I hope ATTAC makes one active. When you are in a political organization it is important how long you have been a member. But in ATTAC [it is different]. Some new members are members of the central organization.

Sharing with us his own experience, Marucs said:

> I am a member of the 15-member group of Sweden. I would like to work at the national level but I would like to give my energy to the local group. If I go for a day to Stockholm, I feel as if I am losing my focus.

During our conversation, Marcus told me that ATTAC did not have money to put up for advertisement, thus news about ATTAC did not percolate in the media. Nevertheless, Marcus told me that once he received a phone call from an old woman who wanted to speak politics with her. For Marcus, transformation lies in such opportunities for intimate conversations. He wants politicians to spend more time with local people rather than networking at the EU capital in Brussels.

Herve Covellec is a friend of Marcus in Malmo and is an activist of ATTAC. He teaches Economics in a local university college. He organizes popular lectures about economic effects of globalization and looks at the work of ATTAC as a popular education movement. So did other volunteers of ATTAC such in Stockholm and Gothenburg.

Malinn Gaewell (2006) has written a doctoral work on the work of ATTAC Sweden and she looks at it from the point of view of activistic entrepreneurship. In this study, she tells us that in 2004, the preferred goals of ATTAC Sweden were expressed as follows:

1   Tobin Tax
2   Abolition of Tax Haven
3   Cancellation of the extended debt of the poorest countries
4   The creation of a fair international trading system which supports poor countries in their effort to develop
5   The creation of a suitable pension system which does not allow speculation with money in pension funds. Pension funds should be used for productivity, socially and ecologically viable investments

Gaewell writes the following about ATTAC Sweden:

ATTAC Sweden is a non-profit voluntary association. But people interviewed underline that ATTAC Sweden is both a network and an organization, it is a social movement and an organization, it is a voluntary solidarity organization, it is a (voluntary) professional lobby organization and a street parliament.

(Gaewell 2006: 132)

In her study, she also writes, "The more radical groups have also showed signs of respect for ATTAC Sweden, even though there have been expressions of considering them as traitors." In terms of the organizational landscape of civic action, some other organizations have developed an appreciation of "developing new forms of organizing" (ibid: 140). Regarding the impact of their participation in ATTAC, she writes:

People who have been involved in the entrepreneurial process of ATTAC Sweden are themselves influenced by this work. . . . Some have left ATTAC work behind them, while some are still engaged in ATTAC work. These people, I argue, are in their other/new positions also influenced by the ATTAC setting and practice in which they have been part of a longer or shorter period of time.

(ibid: 150)

Gaewell tells us that one can become a member of the common working group, the important decision-making body of ATTAC, only for four years. ATTAC Sweden works in the middle of other social movements, civic action groups, and also in the middle of activities. This creates for Gawelli a "suspenseful setting" where actors are required to be more creative. In her words: "People enact their experiences and understandings. But in a suspenseful setting, the variety means that arguments and practices cannot be taken for granted" (ibid: 207).

One of the activists I had met during my fieldwork told me that ATTAC has influenced GATT (General Agreement of Tariff and Trade) discussions in Sweden.

> ATTAC realized the lie within GATT. It has direct consequences in Sweden. Regulating the service sector. Health, water are State-owned but once privatized, they would have to keep on privatizing. In Bolivia, they had struggle over water. They restored the water system.

While working on important global issues such as GATT, this activist tells us that "ATTAC works as a lobby organization." It dialogues with the Government and other political parties. These dialogues create a condition of mutual interaction and appreciation: "They thought that we are dogmatic left-wing extremists. Now ATTAC has proved to them that we are a democratic, knowledgeable organization. When we organized discussion with Government officials, it was a study day for people in the Government." On issues such as this, ATTAC has been able to work with trade unions such as organization of metal workers in Malmo. With such networking and collaboration, ATTAC is able to have a broad cross-class alliance as she says, "ATTAC is middle-class. We are knowledge-oriented. When we take part in demonstrations we come out of our middle-class home."

I had also met with volunteers of ATTAC in Denmark. They were fewer in numbers and were enthusiastic about it during heyday of ATTAC in Scandinavia. There were few members of ATTAC working in universities such as Aalborg where I was a visiting teacher in 2004.

ATTAC in Italy

I had spent a few days with volunteers of ATTAC in Italy in Rome and Genoa in December 2002. In Genoa, Carl Guiliani (March 14, 1978 to July 20, 2001) was shot dead by a police officer during the G7 summit meeting in Genoa in July 2001. This became a symbol of martyrdom for anti-globalization movement. I was visiting Genoa in November 2002 and met with activists of ATTAC there. We visited the memorial of Carl where he was shot dead. We paid our tribute to his soul. Genoa had a few members but they were active and passionate. I visited the home of an activist friend and her mother was not as pessimistic about the then Government of Silvio Berlusconi and neo-liberal moves as himself. She told us that society does not just depend upon a political and economic arrangement and we should learn to look ahead. During our luncheon meeting, she told us how she read the newspaper articles by the philosopher Gianni Vattimo and learnt to appreciate significance of weak thought. Vattimo (1999) speaks about weak ontology where our ontology is not too certain about ourselves. Our mother was challenging us for a new kind of politics which is not too assertive about self and other but genuinely open to transformation from all sources with a dynamics of weak ontology and attendant charity which gives rise to a new kind solidarity.[6] It may be noted here that during my visit to Italy in 2002, I had met with Vattimo in his office in Turin in December 2002. Vattimo, the noted philosopher of Italy and the world who was then a member of European Parliament, received me with inspiring grace and we had a very enriching conversation on the possibility for a new kind of radical politics and openness to the other.

I joined the organizational meeting of ATTAC at Rome on November 26, 2002. Misa was one of the activists of ATTAC. She was a senior citizen. She had retired from her teaching in the schools. She also went out to teach Italian to the immigrants. In her class, she had met with an immigrant boy from Afghanistan. Misha invited Asad, this boy, to come and sleep in her apartment for a few days as he had no place to go except the railway station. Says Misha:

> The condition of immigrants is terrible. The new immigration law is terrible. We need foreigners. My sister who is in wheel chair needs help. No Italian would come and help her! Instead of saying thank you [to the immigrants, the Government and some people are creating problems for them].

During our conversations, Misa told me that ATTAC Italy had collected 150,000 signatures to make Tobin Tax a law, but it could not lead to an enactment of law in the Parliament. Misa had taken part in the European Social Forum (ESF) meeting in Florence in 2002. There were people from all generations in this ESF which was inspiring to her.

Riccardo Liburdi was a leader of ATTAC in Rome whom I had met during my fieldwork with ATTAC Italy in November 2002. He says that ATTAC is an internationalist organization as it works around 54 countries. It follows a horizontal model of organization instead of a vertical one. Speaking about Italian experience, Liburdi says that after the G8 summit in Genoa and the killing of Carl Guiliani, there was a surge of participation in ATTAC. Liburdi says that he and his fellow activists work on creating condition for self-education and mutual education on such crucial issues as economy and environmental problems. As he told me:

> We invite experts to speak to us on issues we do not know. We use internet to organize our meetings. In the West, we live in alienation. We meet once a week. We organize show with the artist to collect money.

But being involved in ATTAC has not been easy for him. He says:

> There is no time. All is quick. I do not have time for myself—to read something for myself. Something for meditation. I cannot do meditation for my salvation. I have an eight hour job and then my work without hour. I do not have a personal life. Therefore, many people do not like politics.

Liburdi's difficulty and frustration with time, life, and politics is shared by many activists. In our communication when I contacted him through Facebook message in January 2006, Liburdi wrote to me that he has totally withdrawn from any public political activity and is now engaged only with his own job and family.

ATTAC Norway

I had visited Oslo in June 2001 and had met with some social activists and volunteers of ATTAC. Monica Sydgard Bernsten was one such volunteer about whom I had heard during my visit. Monica was already there in the G8 summit protest in Gothenburg where I was to come the day after. On my arrival in the G8 summit protest, I met Monica in Gothenburg in a park of protest. She was then writing her master's thesis on ATTAC Norway as a transnational social movement (see Bernsten 2002). In her thesis, she argues

> ATTAC Norway has emerged in the crossroads between the transnational and the local. The transnational ATTAC movement has played a vital part in shaping collective identities within ATTAC Norway. And when members of ATTAC Norway operate on transnational arenas, they become part of a wider transnational community of social movements united in their opposition to neo-liberal globalisation. At the same time the forming of ATTAC Norway is rooted in the hows and whys of Norwegian political context and cultural meanings.

After completing her master's thesis on ATTAC, she started working with ATTAC as an employee working on its media work. Monica says that ATTAC Norway, like all other social action and civic groups, receives support from the Government. Her work is to send information to members on international issues. She compares this with the situation of ATTAC in

France: "ATTAC in France is based upon fight against Government." But in Norway, it is different. Bernsten says that as a worker with ATTAC, she always try to articulate what she is for rather than what she is against.

Bernsten tells us that, in the beginning, ATTAC Norway was a mass movement but slowly it attained an organizational stability. There was tension between the central leadership in Oslo and local groups especially between Oslo and Bergen which reflected the long-standing tension between political and cultural orientations between these two cities in Norway.

I had first visited Bergen in October 2003 to meet with the activists of ATTAC there. Then I joined Bergen Social Forum in 2004 and visited also in 2006, 2007, 2008, and 2010. During my visits, I spoke with activists and participants of ATTAC such as Ase, Espen, Helge, Bhanu, and Elisabeth. Espen was the then leader of ATTAC, Bergen. He was finishing his study of philosophy at University of Bergen. During our conversation, Espen said how ATTAC helps people to go beyond the media-centred view of politics, globalization, economy, and life. ATTAC Bergen, like all ATTAC local groups in Scandinavia and around the world, organizes seminars. Along with initiating spaces for popular education, the local group also helps in local struggles such as young people who are planning to occupy vacant and empty houses in Bergen. In this, participants draw inspiration from the Roofless Movement in Brazil. As the leader of ATTAC Bergen, Espen is for more creative localization of ATTAC. For example, he argued for the radical autonomy of Bergen ATTAC and did not want it to be just a unit of ATTAC Oslo. Espen says:

> A central office has to spend more on letters. People getting salary from Oslo. . . . We are trying to change the tactics of globalization. While starting ATTAC here, we called our friends in Swedish ATTAC and they told us that one can start local branch of ATTAC anywhere. Some people with lot of power in Oslo said: you cannot do it. I thought that it is not ATTAC. I was very angry. But we built our work here. Instead of focusing on what is wrong with Oslo, we started focusing on our work.

Ase Moller Hansen is an engaged activist of ATTAC. She comes from a very diverse background in education, life, and activist engagement. She had studied sociology for her bachelor's degree at University of Bergen. She was a volunteer with organizations such as *Future in Our Hands* and The International Women's League for Peace and Freedom (WILPF). She also took part in international movement against land mining and uranium radiation. She told me that she had also managed a shop and she used it as a place to meet people. Ase wanted to spend more time in creative politics and international peace and justice work, and with the transition in her life such as her divorce, she tried to give more attention to her work for global justice, spiritual work of meditation, and to her own creative writing. Ase was first interested to be an artist. In her words:

I wanted to be an artist. I was drawing cranes. I was spending a lot of energy on small details while the world was out there. I was in *Future in our Hands* for some years. I co-ran a shop. A lot of things were happening in the used store. I wanted to know more about the economy: how money controlled everything. It seemed to be very sick.

Along with my work I also like to write. I write fiction. One of my dreams is to reduce my days of work per week, only to work two days a week and then devote the rest to write.

Ase has been taking a leading role in all the justice and peace works in Bergen. She has been organizing exhibitions for peace and against war. Ase draws inspiration from peace activists and researchers such as Johan Galtung, Norwegian philosopher Arnae Naess, and Norwegian politician Gunnar Garbo who once told in an exhibition that war is terror. He also draws inspiration from Norwegian writer Jens Bjornebo (1920–1975) who was a critic of Norwegian society and Western civilization and wrote a novel called *Moments of Freedom*. Ase wanted a creative space to bring herself and political struggle for peace and justice together and ATTAC provided this to her. As part of her creative self-expression, Ase also writes novels depicting the condition of Muslim immigrants and refugees in Norway. Now she devotes much of her time to her work with The International Women's League for Peace and Freedom (WILPF).

Helge was another activist of ATTAC Bergen that I met. He was then completing his PhD in Economics from University of Bergen. Helge has been deeply interested in environmental issues. He says, "Once you have children, you become interested in ecological issues." Helge also writes for children on such social issues as racism. He says that he writes theses in a way which is fun.

While doing his studies at the university, Helge became drawn into antiglobalization movement. He took part in the protests against WTO meeting in Prague in 2000. He had not heard about ATTAC then. He read about ATTAC from a newspaper article in Norway and then he joined in. Helge, like Ase, brings action and meditation together in his engagement with ATTAC and global justice movement. Helge has been part of a Buddhist meditation group.

ATTAC was active in many countries and I had met with activists of ATTAC in Budapest and Brazil as well. I had visited Budapest in December 2002 and had met with and spoken to activists of ATTAC there. ATTAC provided a creative and critical space for engagement to some of them, as they were struggling to come out of previous regime of state socialism and were feeling crushed by the pregnant march of neo-liberal global capitalism.

Camila Hansen is another activist of ATTAC whom I had met during my fieldwork in Norway and I am in communication with her. Camila is visually challenged and she is devoted to the causes of ATTAC such as global tax justice and democratization. Since November 2019, she has been working as a political advisor for ATTAC Norway and writes on trade and environmental issues. In a recent email communication, she writes about her work and research:

> Since November 2019 I've been working as a political advisor for Attac Norway, specializing in trade politics. The job is a trainee position which will last until May this year. What I work with is the impact of trade agreements on the environment, human rights and democracy. I write articles as well as analyses of trade agreement texts. Some of my texts are published on Attac's website (https://attac.no/author/cami78/), but they are all in Norwegian. Currently I'm writing on a policy brief about the impacts of neoliberal trade and investment agreements on the environment and on climate change, and proposals for alternative trade policies.

I had met with some of the volunteers of ATTAC in Brazil. In Brazil and Hungary, ATTAC was small but members had links to other social movements. Members of ATTAC in all these countries also had links to and sometimes active participation in different social forums, World Social Forum as well as different local, regional, and national forums.

Taxation of currency transactions (Tobin Tax) and just taxation of multinational companies as part of a broader movement of global justice are some of the main objectives of ATTAC. Some of the recent moves seem to support these. For example, at the latest G7 meeting on June 5, 2021 in London, the finance ministers and central bank governors of G7 countries came to agreement over two proposals to counter international tax avoidance.

- The first measure is to mandate companies to pay taxes in the area where they conduct business.
- The second measure is to enforce a global minimum corporate tax rate of 15% to counter the presence of low-tax jurisdictions.

Social Forum Movement and the Calling of Global Responsibility

The project of society is not an extension of either state or market. Social project is a project of creating self and society anew, and it is a project of creating a new social becoming.[7] The Social Forum Movement tries to realize such a path of social becoming. The Social Forum project began with World Social Forum in Porto Alegre in 2001. It was held in Porto Alegre, Brazil, for many years. But then it moved to Mumbai in 2004, and from then on it has been around in different parts of the world. The last WSF meet was held in Tunis in 2015 and the next one is taking place in Montreal.

WSF has built upon alternative traditions of thinking about globalization such as mundialization from France (see Smith et al. 2014). It resonates with the spirit of a thinker like Jean-Luc Nancy who would challenge us to realize that the present-day globalization as neo-liberal globalization is creating a world which is not habitable, an *unworld*. In its place, we need an alternative process of "mondialization" which counters this and creates our world as a place of mutual habitation (Nancy 2007).[8] It also resonates with the vision and practice of a thinker such as Chitta Ranjan Das (2008) who also challenges us to build people's power as a foundation for a genuine globalization rather than a spurious one.

WSF has created a transnational global space for many different kinds of actors and organizations to meet. According to Boaventuara de Sousa Santos (2005), it is a place of subaltern cosmopolitanism where subaltern social groups with alternative visions and social projects come together. de Sousa Santos is a creative sociologist from Portugal who has been active in WSF movement from the beginning. According to de Sousa Santos, WSF gives rise to critical realist utopias in our times, as it fights against hegemonic globalization, neo-liberal regimes of governance and legality, and social fascism at home and the world. WSF provides a space where different individuals, groups, and social movements come together. They argue with each other, sometimes are in conflict with each other, but nonetheless there emerges a trans-conflictual field of communication and emergent and moving commonality. As de Sousa Santos tells us about it:

> For the movements and organizations in general, what unites has been more important than what divides . . . even when cleavages are acknowledged, the different movements and organizations distribute themselves among them in a non-linear way. If a given movement opposes another in a given cleavage, it may well be on the same side on another cleavage. Thus, the different strategic alliances or common actions featured by each movement tend to have different partners. In this way are precluded the accumulation and strengthening of divergences that could result from the alignment of the movement in multiple cleavages. On the contrary, the cleavages end up neutralizing one another. In such trans-conflictuality, to my mind lies the WSF's aggregating power.
>
> (de Sousa Santos 2005: 33)

de Sousa Santos (2008) tells us that WSF strives to be an open global space, a space of movement of movements. These movements are often grounded in "multi-secular cultural and historical identities" (2008: 10).

> It has created a meeting ground for the most diverse movements and organizations, coming from the most disparate locations in the planet. . . . Some are anchored in non-Western philosophies and knowledges that sponsor different conceptions of human dignity and call for a variety of other worlds that should be possible.
>
> (ibid: 11)

Furthermore, "WSF represents the maximum possible consciousness of our time. Its weakness—the inability to discriminate among diverse solutions—cannot be separated from its strength—the celebration of diversity as value in itself—and vice-versa" (2008: 12). de Sousa Santos also writes:

> The WSF signifies the re-emergence of a critical utopia, that it is to say, the radical critique of present-day reality and the aspiration to a better society. The WSF challenges the total control claimed by neo-liberalism, whether knowledge or power, as a way to affirm the possibility of alternatives.
>
> (ibid: 18)

de Sousa Santos further tells us:

> The utopia of the WSF is a radically democratic utopia. . . . It helps to maximize what unites and minimize what divides, it celebrates communication rather than disputes over power, and it emphasizes a strong presence rather than a strong agenda.
>
> (ibid: 18)

While focusing on immediate tasks, it also challenges us for a civilizational paradigm shift in terms of approaches to power, not just acquiring power but to "transform power" (ibid: 18). WSF tries to address "unresolved tension between contradictory temporalities. . . . Calls for immediate debt cancellation get articulated with long duration campaigns of popular education concerning HIV/AIDS" (ibid: 19). WSF has "mobilized pragmatism in combination with the reconceptualization of diversity as a strength to produce energy and political creativity" (ibid: 28). "The WSF has created a political environment in which politicization may occur by means of depolarization" (ibid: 29). "The WSF underlines the idea that the world is an inexhaustible totality, as it holds many totalities, all of them partial." For de Sousa Santos, it also enacts a passage from a "movement politics to an inter-movement politics" (ibid: 44).

As there are annual WSF meetings, there are also regional Social Forum meetings such as Asian Social Forum and European Social Forum. There are also national forums and regional forums such as Danish Social Forum, Indian Social Forum, and Scanna Social Forum (regional forum in Sweden bringing together and activists from the Southern part of Sweden called Scanna which includes cities such as Lund and Malmo).

I had taken part in the World Social Forum in Mumbai in 2004. It was a festive occasion. What struck many of the people who had already taken part in WSF meets in Porto Alegre (Brazil) is that it brought a much more diverse range of participants from different marginalized and struggling communities. In WSF at Mumbai, there was a larger presentation of Dalits and women. WSF Mumbai was also a place for bringing together social activists and peace workers from India and Pakistan. WSF provided a post-national and trans-national context for activists of India and Pakistan to speak about struggle for peace in Kashmir. For example, one of the activists from Pakistan in a session on Kashmir commented: "The Kashmir issue cannot be solved in the framework of nation-state. We need a post-national framework."

WSF was also a place for bringing together many social movements from around the world including activists of ATTAC. It had also brought activists of Ekta Parishad to this World Forum. But Ekta Parishad had organized a meeting in a separate venue and had organized a festival for struggle for land rights called Land Mela. Five hundred tribal people from around the country as well as many co-participants in Ekta struggle from Europe such as the noted thinker Dr. Karl-Julius Reubke had taken part in this. While for Dr. Reubke, the dominant slogans in the main WSF venue were much more anti-American as in Arundhati Roy's inaugural speech at the WSF denouncing the evil regime of George W. Bush, in the land mela of the tribals, it was much more an occasion of festive determination of positive struggles on the ground. It was a struggle to get land to live rather than be just a nomad at home and world in the days of neo-liberal global capitalism. In the land rights festival organized by Ekta Parishad at WSF Mumbai, there were also inter-linked meditations on the state of the world. Christopher Strawe, a devoted participant in the path of Rudolf Steiner and his vision of creative world transformation, spoke about "Threefolding and Global Governance" in which he proposed Steiner's pathways of threefold social life where market, culture, and state have their own autonomies rather than either of these playing an annihilating dominant role (cf. Steiner 1985). Such a vision and practice goes against the ideology of either state or market fundamentalism or communitarian or culturalist fundamentalism which leads to production of poverty, misery, suffering, and annihilation of potential of self, culture, society, and the world. Nicanor Perlas, the noted activist and friend from the Philippines who has also been engaged in many projects and social movements, had taken part in this. Perlas (2003) argued how we need to fight for an alternative globalization based upon people's power.[9]

In WSF Mumbai, Reubke met many Dalit activists. During their conversation, many of them shared their critique of Gandhi and preference for Ambedkar. They told him that they want education. But then when Reubke asked them what kind of education they want, they could not elaborate it further. Reubke comments, "What kind of education they asked for. This seemed a funny question since everybody knows what education is. I understood that for them it was a mysterious black box probably full of gold" (Reubke 2020: 359).[10]

The Social Forum movement has spread across many different countries and regions of the world. Many countries like Denmark and Brazil organized annual Social Forums in the first decade of 21st century. I had taken part in Danish Social Forum in October 2003 and Brazilian Social Forum in

November 2003. In Danish Social Forum, there were many sessions including one session led by Dada Mahesvarananda of Ananda Marga on his book *After Capitalism*. In Brazilian Social Forum, there was a session on "Postliberalism." Both the social forums brought also many religious and spiritual groups together such as Ananda Marga. World Social Forum thus has provided a much more open space to faith groups and is part of a much-needed post-secular spiritual transformation of humanity.[11] But faith groups are now coming together in Parliament of World Religions in much more creative and celebratory manner. Many of the participants there embody struggle for global justice along with struggle for dialogues (more on Parliament of World Religions in our next chapter).

In his study of World Social Forum and related alter-globalization movements, Geoffrey Pleyers tells us that in World Social Forums as well as local, national, and regional forums, actors express themselves as well as bring their expertise what Pleyers calls ways of subjectivity and ways of reason. In World Social Forum in Mumbai in 2004, there was a balanced combination of both these aspects—ways of subjectivity and ways of reason. Here what Pleyers writes deserves our careful consideration:

> They constitute open spaces . . . intended to foster exchange and collaboration among actors belonging to the two alter-globalization trends. Conceptualized by committed intellectuals strongly tied to the way of reason and rather institutionalized, the very first WSF was soon overwhelmed by the creativity of participants from all over the world, the mix of experiences, autonomous spaces and the increasing will to implement alternatives. Most activists consider the informal exchange of activist experience with people from different countries, during attendance at official forum workshops, to be a privilege. The Indian WSF reinforced this expressive aspect with its dances, songs and theatrical performances of the crowd in the alley-ways of the Forum. It also underlined the importance of concrete alternative implementation, including the use of free software. The meeting between the two paths also played out at the level of each individual participant, who shuttled constantly from a technical workshop to the swell of subjectivities in the alley-ways and the exchange of experience with other activists. The Social Forums are thus simultaneously spaces for the elaboration of expertise and for experimentation with another world; for popular education and for exchanges of experience.
>
> (Pleyers 2010: 183)

WSF started in Porto Alegre in 2001, and in 2021 it completes 20 years. Now there is a move to change WSF from a place of meetings to a place of action. Some of the original founders of WSF now want to change it to a place of action so that WSF can play an effective role in world change. In the beginning, the Forum wanted to avoid conflicts and make WSF a place for conviviality and consensus. But arriving at consensus was difficult and WSF was not represented in many critical movements of our times such as movements for climate change, Black Lives Matter, Me Too, and the Arab Spring. Now the founders want to make it a global political agent. Some of the founding members of WSF such as Boaventuara de Sousa Santos and Vandana Shiva wrote in a letter:

> It is crucial to introduce a governance that allows us to go from an open space to a space of action. We therefore propose that the International Council integrates the new social forces that are mobilizing all over the world, that we try to make a more representative body which can take a look at our Charter of Principles so as to adapt it to the new times of the 21st century, that we give a place to the regional and thematic forums, that we organise international days of action, that we can discuss the road to follow to make the WSF a global political subject. This will not be an easy road and will require the openness and willingness of all of us in order to create an efficient body able to speak to the world. There are many possibilities to strengthen the governance and to democratize the WSF. We very much hope they will come out so they can be democratically discussed.

At the same time, some of other founding and nurturing participants of WSF such as Gillio Brunnelli of CIDSE, Canada, feel that it is important for WSF to be an open space where organizations and movements around the world can meet and share their experiences. For Brunneli,

> [I]t must also be recognised that if a broad and open space like the WSF had not existed and did not continue to exist, these campaigns/movements would not have had the global importance they have had and would probably have remained important social phenomena, albeit localised. It is because, over the years, the WSF has contributed to building networks of organisations and associations, to opening channels of exchange and communication, to bringing together civil society leaders from all over the world, that local struggles and specific issues can be relayed and supported worldwide almost immediately without using traditional media.

On its 20th anniversary, World Social Forum was held online from January 23 to January 31, 2021, and this year, it focused on climate change/ecology, peace and war, democracy, social justice, economic justice, etc. It began on January 23 with a world march for justice, cultural political act, democracy, and well-being and local, face-to-face, and virtual activities in different parts of the world. The next edition of World Social Forum was held on Mexico City in person from May 1 to May 6, 2022 where participants focused on the challenge of building a different world differently.

Indignados

Indignados has been another social movement in Europe in recent years which has expressed indignation of common people towards inequality and injustice created by neo-liberal policies and globalization. It was active two years ago in France, Spain, and Portugal and other countries. In some countries such as Spain and Portugal, it continues its active mobilization. For example in Spain, Indignados has recently won some mayoral and local elections.

Indignados drew inspiration from the life and thought of Stephane Hessel, an inspiring fighter of resistance against Nazi occupation of Europe. Hessel, originally coming from France, was part of resistance against the Nazis during World War II. As part of his resistance work, he had moved to London but he came back to France to fight against the Nazis and was captured. He only narrowly escaped death in the concentration camp in Buchenwald. In an interview, Hessel tells us: "You must find the things that you will not accept, that will outrage you. And these things, you must be able to fight against nonviolently, peacefully, but determinedly." "You must find the things that you will not accept, that will outrage you. And these things, you must be able to fight against nonviolently, peacefully, but determinedly." About his life, Hessel says:

> I was then sent—unfortunately, I was captured, sent to Buchenwald. And when the war was ended, I said, "Now I have a responsibility, as a survivor," as one of the relatively few survivors of concentration camps—many of my comrades had died there. So I said, "Now I have a responsibility. I want to carry it." What is the responsibility? It is to let the values, on which we have fought, known to the succeeding generations. That is what I've been doing until now.

After the war, Hessel helped draft the United Nations Universal Declaration of Human Rights. He was also part of Ethical International Collegium. Witnessing the widespread of human suffering and misery created in the recent times especially after the American financial crisis and the ensuing European financial crises, Hassel felt compelled to issue a manifesto of outrage against this. He wrote a pamphlet, *Indignez-vous, A Time for Outrage.* *A Time for Outrage* is a 35-page book written by Hessel in 2010 which sold three million copies in 30 languages and inspired protests like Occupy the Wall Street in the USA and Indignados in Spain.

In his book, Hessel says, "the worst attitude is indifference." He wanted to transform this by writings this book. As he says:

> I was worried that so many young people in all our countries seem to have forgotten their responsibility for values. They are just responsible to find a flat, to get some money, to have material wealth. And they do not realize that that is going to be jeopardized if the basic democratic

values are not fought for. And that is where I think this indifference, which is widespread in many of our countries, obviously—resistance has always been a minority act, and we need minorities, and then they will spread. But indifference, just let it be, all discouragement—we wanted to do something, but it failed, and we are no longer capable of doing it. That is the danger that I try to fight by telling young people, "Have confidence. Trust your strength. If you go to the streets in a determined way, you will see the government will have to listen to you. And you have to be confident and brave."[12]

Hessel inspired movements such as Indignados. Drawing inspiration from Hessel, Indignados movement expressed resistance against indifference and was a struggle for realization of responsibility. Indignados movement sprang in Spain, Portugal, and other countries. Manuel Castells, the noted sociologist of movements, writes about Indignados:

> On December 19th past, after an assembly discussion, the Madrid Puerta del Sol Commission for International Outreach decided to suspend its activity and declared itself on indefinite active reflection. "The public space we had rediscovered has been replaced once again by a sum of private spaces. . . . The success of the movement depends on us being the 99% once again. Although we do not have the answer to what has to come next, what shape the restart we need can take, we understand that the first step for escaping from the wrong dynamic is to break with it: to stop, hold back, and get perspective", went the argument.[13]

Indignados fought against housing closure in Spain. It built upon a democratic movement called Democracy Real. It also stressed on alternative economic practices, consumer co-operatives, ethical banking, and exchange networks. In local municipality elections in Spain, Indignados has won in some places. I had visited Spain in January 2013 and had met with a few Indignados activists. In the meantime, Indignados has grown to be an important political force in Spain and has grown into two political parties playing a major role in Spanish politics.

The Indignados movement in Spain was the beginning of a phase of drastic changes in Spanish society.[14] The movement that began as a mass, citizen-led response to the country's regime crisis continues to present its effects through the two political parties that arose from the movement: Podemos and Ciudadanos. The roots of the movement can be traced back to the 2008 recession that drastically affected Spain. Mariano Rajoy of Partido Popular (PP), the then Prime Minister of Spain, brought in austerity measures to tackle the crumbling economy. These measures introduced by the government "reduced democracy to the realm of finance" (Antentas 2017: 107). However, he argues that it was not just austerity measures that prompted mass mobilization into the streets but a gradual shift towards neoliberalism over 30 years since

Spain's transition from dictatorship to democracy in 1975. The rise of neo-liberalism resulted in what Antentas terms a "process of de-democratization." Politics and representation became subordinated to financial interests, which disappointingly prioritized emergence in an advantageous position from an economic crisis rather than society in itself (ibid: 107).

Subsequently, the problem in Spain cannot be solely viewed as a financial crisis. The very idea and representation of Spain, as known by the world, had collapsed. The Marca España or the Spanish Brand that associated democracy with welfare was eroded. All the political and economic projects that were developed by Spain post transition into democracy also experienced a crisis. Post transition, when Spain had joined the European Union (EU), it saw dramatic growth in its economy and recorded high levels of satisfaction from the Spanish public. It was often seen as a success story in the European context. However, the austerity measures from the same Union pushed the nation to the shores of a crisis. Antennas related this collapse to the shift of the idea of progress to "capitalist modernization strategy and the expansion of consumerist society." This reality came with a worsening living conditions for the working class and decay in the standard of living for the middle class (ibid: 111).

Therefore, the Indignados movement that burst out into the streets on May 15, 2011 was a twofold criticism of the Spanish state. (They were called 15 M demonstrators.) First, it criticized the economic and financial powers as being responsible for the recession, and second, it criticized the *clase politica* or the political elite for being complicit with the business world. The movement rose from the margins of the economy to the centre of the Spanish socio-economy, as the Left had failed to do their part and led the country against austerity measures. This brought about the slogan "No nos representan," which translates to "they do not represent us," proof of a political system that was hijacked by financial power (ibid: 111). Thus, many authors invoke Gramsci's term of a "regime crisis" in Spain. This refers to a crisis where the nation experienced a total failure of the institutions codified in its constitution, in addition to problems of political representation. Furthermore, the Spanish case could also be interpreted as Gramsci's crisis of hegemony of the Spanish elites who were involved in the mishandling of the economic crisis (ibid: 107).

The 15M demonstrations arising from the regime crisis was a turning point in Spanish history, and it created a clear before and after timeline. Antentas considers the demonstration to be not just a single mass movement but the beginning of the movement towards re-politicization (ibid: 112). It shook up various communities within Spanish society, the two main political parties that controlled Spanish parliament, Partido Socialista Obrero Espanol (PSOE), and Partido Popular (PP) and spread across different parts of the world as square occupation movements. The scale of the rebellion was such that it was considered most important after the resistance under the Franco dictatorship that led to the democratization of Spain in the 1970s.

It challenged the very foundations of the existing political system and institutional framework that were dominated by PP and PSOE. This movement was also significant in the constituents of its participants. Supporters cut across all generations of Spanish society. Notably, the majority of its participants were the youth who were born into a defected system that was riddled with unemployment and uncertainties. On the other hand, immigrants also made up an important part of the 15M protests. However, they were an invisible part of mobilizations, reflective of their exclusion not only from Spanish politics but also from Spanish society. It is also valuable to note the social response to the Indignados movement. In a survey conducted in June 2011, about a month after the rebellion broke out into the streets, around 80% of the country's population believed that the movement was justified in their aims (ibid: 113).

The scale of the movement can also be measured in relation to the monumental changes it brought along. Diverse initiatives and projects were organized with their own set of aims, while interconnected with one and other. The positive outcome of the 15M was a re-politicized Spanish society that was more polarized than before, taking active interests in collective affairs thereby drastically changing the social climate. Unfortunately, austerity measures continued to hammer the lives of Spaniards, leading to a bleak future (ibid: 114). Antentas elucidates two major successes that stemmed from the movement, the formation of Plataforma de Afectados por las Hipotecas (PAH) and the mobilization on privatization with an increased focus on education and healthcare. These initiatives had far-reaching effects that sustained over long periods of time via different organizations. These were termed as "tides." The PAH was a collective of the families who were unable to pay the mortgage on their homes and were facing the threat of evictions. This movement was the extension of the right to housing that was championed in 2006 under the banner of V de Vivienda. On the other hand, these "tides" engulfed different public sectors which were associated with specific colours (ibid: 115). The most important were the healthcare and education sectors, identified with white and green, respectively. However, the most striking result was the rise of political parties that shook up the traditional party system that was previously a limited choice between PP and PSOE. The political phase of the regime crisis was struck open, exposing the weak institutions of the country.

Podemos began as a collective of intellectuals and activists who were led by Pablo Iglesias. They began closely related to the 15M movement but later diverged to form a left-wing populist project that was widely recognized for its new style by raising concepts and issues into the national political debate. The change it brought about to the political spectrum was so drastic according to Antentas that political life experienced a "podemization" process (ibid: 124). Although scholars categorize the party to be in the left wing, the leaders have strived to maintain an image of not their political leaders or cadres but as an instrument of political change (Iglesias 2015: 19). Sola and Rendueles, in "Podemos, The Upheaval of Spanish Politics and the Challenge of Populism," write that the party never referenced the right–left wing political spectrum to define their position. They never invoked the measure to place themselves in relation to the other contending political parties. This intentional move was quickly defeated with the rise of right-wing party Ciudadanos. The party rose in opposition to increasing Catalan nationalism while benefiting from the wide change brought in by Podemos. The party championed itself as advocates of sensible and calm change in comparison to a radical Podemos. They stressed on a market-centred programme and had backing of most media houses. Soon they were perceived and labelled as the "Podemos of the right" (Sola and Rendueles 2018: 111). Ciudadanos also contributed to the end of the "bipartyism," as there were now four main political contenders: PP, PSOE, Podemos, and Ciudadanos.

The proof of this lies in the results of elections beginning from the European elections of May 25, 2014 (see Simón 2020). PP and PSOE together won only 49% of the vote compared to the 81% in the previous year's elections. This was attributed to the rise of Podemos who managed to win about 7.9% of the vote thanks to its roots in the 15M movement. Following this, in the local and regional elections in May 2015, the new parties continued to show results although the effect of their freshness had declined. Podemos trailed behind both PP and PSOE, while Ciudadanos was even behind Podemos (Antentas 2017: 124). In the same year, the December ballot showed similar trends. The bipartisanship crisis continued when the support for Podemos slowly deteriorated. PP and PSOE obtained 50.7% of the votes as opposed to 73% in 2011 and 83.8% in 2008, a drastic fall. Meanwhile, Podemos won 20.6% of the vote, very close to PSOE (22%) and not so far behind PP (28.7%) and Ciudadanos remained far behind (13.93%) (ibid: 125).

Pablo Iglesias, the face of the anti-austerity party Podemos, writes that the core members of the party, fellow lecturers at the Complutense University, never expected the party to reach the heights that it has. They began by merely focusing on actions such as writing papers, promoting small-scale initiatives, producing and presenting TV programmes, studying audiovisual communication, and advising political leaders on media strategy (Iglesias 2015: 14). What resulted from these endeavours was the high recognition the party began to have in the country that worked towards its favour in the future elections. The novel hypothesis of Podemos, he writes, rises from their personal political experiences as well as a model for political communication. The core group of members were "radicalized in the nineties, amid swirling currents of Negrian [referring to political theory of Antonio Negri] political theory and alter-globalism and their presentational skills were first honed on community TV" (ibid: 7). Analysing the regime crisis through the lens of political change in Latin America gave them the tools to interpret the problem within the context of the Eurozone. They began initiating the "latinamericanization" of Southern Europe as opening a new structure of political opportunity (ibid: 14). The other key factor to their hypothesis was the creation of a TV nation through their televised pro-gramme La Tuerka, meaning "the screw." The power of the media not only enabled the party to enter into the homes of Spaniards nation-wide but also taught them important skills and tools to further their political conquest. Iglesias describes that by operating on the "fundamental terrain of ideologi-cal production," La Tuerka trained them to intervene into the mainstream debate, perform consultancy work in political communication which in turn led them to gain experiences in electoral campaigning and advising political leaders (ibid: 14).

It was a combination of the 15M movement, activists, and lecturers and the strength of television that empowered Podemos to build the narrative of a certain "caste" in Spain. They distinguished two communities: the people and the caste, positioning them at opposite sides of the spectrum. The caste most often referred to politicians, bankers, speculators, and any other privi-leged group to whom anyone could express indignation against the establish-ment (Sola and Rendueles 2018: 104). This term resonated with the Spanish population and soon entered everyday language. An interesting aspect of the party is the demographic of its members and voters. While the vote is highest among young, educated, urban dwellers, and men, Sola and Rendueles note that this is paradoxical as Podemos' ideals are akin to feminism. The party has the highest percentage of women among its national deputies which might possibly relate to the fact that women have been known to vote for PP and PSOE in the past (ibid: 109).

Present-day Spanish politics is heavily influenced by the 15M protests and the roles of Podemos and Ciudadanos. While the parties' electoral gains are seeing recorded lows, the Spanish political spectrum is highly divided. Spain saw two general elections in 2019 in the months of April and later in November as no conclusive alliance was formed between the different political parties. In the April 2019 election, the turnout was noted to be over nine points higher than the 2016 election. PSOE obtained 2.5 points and 2 million voters more than in 2016. It won 123 seats, more than double the number of seats won by the party that came second. On the other hand, PP lost 71 MPs and more than 3.5 million voters. It was able to clinch only 66 seats owing to the corruption scandal the party was involved in. Next, Ciudadanos gained 2.8 points, only one point away from overtaking the PP's position, close to being the right-wing party with the largest vote gain. Unidos Podemos, the alliance of Podemos and other left-wing parties like the Izquierda Unida (IU), lost 29 seats (almost 7 points), and one-third of its voters. The game changer in Spanish politics was the entry and rise of the far right party, Vox, which emerged with 2,677,173 votes, 20.2%, and 24 seats (Simón 2020: 19). Ultimately, PSOE did not win a clear majority to form the government. An alliance needed to be created in order to form a coalition government. Despite multiple negotiations with Unidas Podemos, the two parties failed to come to a consensus. This failure eventually led to the second general election, almost six months later in November.

In the span of half a year, the results of the election showed drastic changes. PSOE continued to win the largest number of seats, although insufficient to form a majority government. The other left-wing party, Unidas Podemos, managed to retain only 35 seats of the previously won 42. Transformative changes can be noted in the results of right-wing parties. While PP managed a comeback and gained 27 seats more, Ciudadanos saw a crushing defeat. It won 55 seats in April but managed a win of merely ten in November, possibly seeing the beginning of its end. This defeat can be attributed to the rise of Vox, and the party which previously did not have much representation in the government is now the third largest force in the Spanish parliament with 52 seats, following PSOE and PP (País 2019). Yet again Spain saw the dire need of a coalition government in order to progress. PSOE led by Acting Prime Minister Pedro Sánchez stated that a progressive coalition government which was a good opportunity in April was now a dire necessity. Finally, a minority coalition government was formed between PSOE and Unidas Podemos which was sworn in with the approval of other left-wing independent parties. This is a notable stage in Spain as it is the first minority coalition government in the modern history of Spain since the formation of the Second Republic (País 2020).

The Indignados movement has been the seed for the growth of two major political parties. What began as a mass mobilization led to the unison of lectures and activists who launched Podemos, whose general secretary, Pablo Igelsias, now leads as one of four deputy Prime Ministers in the current Spanish government. The regime crisis, stemming from political and financial failure, shook the entire nation. Consequently, the new political parties continued to stir Spanish society with not only their policies but also their mere presence. While Ciudadnos and Podemos are recording lower electoral gains each year, they're effect on Spanish politics is as powerful as before. The proof of their impact lies in the first ever modern coalition government. The current Spanish society has diverse and differing political opinions than ever before, a drastic change from the routine dual choices of PP and PSOE.

Immigration Policies of Ciudadanos and Podemos

Spain plays a key role in immigration within the European context due to its strategic geographic location along the Mediterranean, close to the African subcontinent. In recent years, Spain has been receiving a large number of immigrants, especially asylum seekers. In 2019, Spain received 18,264 Applications, a whopping 112% increase from the year before. On the other hand, decisions were taken only on 60,198 applications, leaving behind numerous pending cases (UNHCR 2020). Hence, immigration policy became an important part of the public narrative. Votes in the last few general elections have been heavily influenced by the parties' stance on immigration.

In their party manifesto for the November 2019 general elections, Podemos's proposals are focused on providing safe migratory passages with a heavy stress on the protection of human rights. While PP, PSOE, and Ciudadanos have a rather similar approaches, stressing on orderly migration by reducing illegal channels, Podemos focuses on bettering the existing system. Their proposals include the easing of regulations of family regroupment, allowing asylum claims at Spanish embassies abroad and other third party nations and creation of new types of visas such as employment, and importantly for humanitarian reasons (Podemos 2019: 63). Recognizing the loopholes in the existing asylum system, they propose reformation of the system such that there are no forms of racism and strive to include climate refugees, victims of human trafficking, and the LGBTQ+ community (ibid: 64). Their aim has transformed into their goal of ensuring that there are zero deaths in the Mediterranean migratory sea route in 2020. In addition, they also promise to work with other stakeholders to prevent the controversial "devoluciones en caliente" or hot returns where migrants who reach the Spanish border are deported without a chance to explain their circumstances (ibid: 63). Additionally, in a rather reforming move, they propose to abolish CIEs (Centros de Internamiento de Extranjeros) or foreign detention centres that they claim are complete violations of human rights (ibid). Finally, they stress on the intersectionality of the vulnerability of migrants (ibid: 93). Migrants are also exposed to other obstacles such as homelessness which risk their health and dignity of their lives. Hence, the left-wing party displays a humanistic approach to immigration focusing on the all-round challenges they face. Podemos particularly pledges responsibility to ensure their human rights in accordance with international agreements.

On the other hand, contrary to Podemos' robust position on the matter, Ciudadnos presents a rather centrist approach, similar to PP. Their goal is to achieve orderly migration routes that remove irregular or illegal migration into the country. They thus advocate for the categorization of human trafficking as a crime in the civil code as it profits from risking the lives of migrants (Varela 2019). They also propose better patrolling and border control to ensure that there are no irregular border crossings. This can also be observed in their election manifesto of the 2016 general elections. Ciudadanos focused on the creation and implementation of a cohesive border control plan that is uniform throughout the European Union, especially to manage irregular economic migration (Ciudadanos 2015: 44). They further propose to support this system with adequate financial and human resources. This proposal arises in order to reduce the burden of immigration on countries at the border such as Spain, Italy, and Greece. While proposing better integration of migrants into Spanish society, they also focus on bettering processes to allow for the voluntary return of migrants and the introduction of visas to attract special talent from abroad (Varela 2019).

Therefore, a humanistic approach to fix the existing system in order to safeguard immigrant lives can be observed with Podemos while Ciudadanos offers a rather neutral lens. Ciudadnos' manifesto not only contains proposals to better the experience for existing migrants but also takes sufficient charge at the national borders to prevent irregular means of crossing.

Occupy Movements

Ever since the onset of neo-liberal capitalist globalization and onset of economic crises beginning with Asia and the subsequent protests at WTO meetings beginning with Seattle, the visions such as "People, Not For Profit; The World is Not for Sale" and "We are 99%" inspired millions across the world to fight for a different world which is more just and beautiful. After the American financial crisis of 2008 and the subsequent European financial crises, people started realizing how their lives are being controlled by financial capital and how politics and state have become a slave to it. They started expressing their rage against it. Movements like Indignados were an expression of it. People also started occupy movements, occupying the Wall Street of New York City and other cities and business centres all across North America and Europe. Some scholars and activists such as Arjun Appadurai started talking about Occupy Economics Movement as a way of transforming the very foundation of current economic order which also raises the issue of not only an ethics of probability but also an ethics of possibility.

On September 17, 2011, around 2,000 people filled New York's Zuccotti Park in New York City. A few hundreds stayed that night, and the night after. They established tents and stayed there. They declared, "We are the 99%." The authorities tried to disperse them by hook or by crook but they did not succeed. Rather the crowd swelled to some tens of thousands. Soon Occupy city movement spread to nearly 500 cities in the USA until it was forcibly cleared by the authorities.

Anthropologist and anarchist thinker David Graeber had played an important role in organizing and articulating the aspirations of this movement in New York City and elsewhere. With his works like *Debt: The First 5,000 Years*, he had given voice to the emergent energizing slogan, "We are the 99%" (Graeber 2011). In his book *The Democracy Project: The History, a Crisis, a Movement*, Graeber (2013) writes about this movement as part of a long struggle for democracy and recent struggle against neo-liberal market capitalism. Graeber reflects upon the challenge of organization in such movements and how the creative organization of these can pre-figure an aspired for society of freedom that the movement is fighting for. For Graeber, in the Occupy movement, one finds a "pre-figurative politics." As Graeber tells us: "the idea that organizational form that an activist group takes should embody the kind of society we wish to create." For Graeber, the occupiers created spaces of freedom where democratic discussion and problem-solving became possible. At the same time, there came up some structures of exclusion during the occupation of the park. The assembly decided that the homeless people of the city could not be served with food and shelter during the occupation.

Movements like Occupy formed alter-globalization movement which had precedents in mobilizations such as ATTAC and protest against WTO meet in Seattle in 1999. As discussed earlier, Geoffey Pleyers (2010) has studied such movements in *Alter Globalization: Becoming Actors in a Global Age* and tell us how all these movements struggle for a new relationship between state and society and respond to demands of justice—local as well as global going beyond the dictates of the market and logic of profit maximization of neo-liberal capitalism. As the noted sociologist Alain Touraine writes in the preface to this book:

> The alter globalization movement asserted the necessity of breaking with the Washington Consensus and seeking the equivalent of what was, after the Second World War, alliance between a strong state and social movements, sufficiently powerful to push the state to subject the economy once again to the demands of justice.
>
> (Touraine in Pleyers 2010: xvi)

During my visit to New York City in March 2013, I met a man from Italy who had taken part in the Occupy Wall Street movement. He told me how he felt energized by it. He felt a larger sense of purpose of life and commitment to the world. He had left Italy for the USA and he does not have a regular job. But being in the Occupy movement, he could feel the pain of the larger world. In a limited way, it is individuals like him who embody an ethics of possibility that Appadurai (2013) talks about.

Ekta Parishad

Ekta Parishad is a social movement cum voluntary organization in India which has been working for the last 25 years with the tribals and other marginalized individuals and societies. Ekta Parishad draws inspiration from Gandhi in its work with tribals and other marginalized peoples. Tribals are one of the most neglected and marginalized sections of Indian society and scholars such as Ramachandra Guha attribute to this lack of concern for tribals in the nationalist struggle for freedom and in Gandhi. As Guha writes:

> Adivasis are far more vulnerable than Dalits . . . or Muslims, and yet their vulnerability is discussed far less in political and policy debates. The reasons for this go back to Gandhi. . . . Gandhi said very little about the Adivasis.
>
> (Guha 2015: 882)

But Ekta Parishad drawing inspiration from Gandhi is striving to identify with the suffering of tribals in a way creating new possibilities for Gandhian vision and pathways. Ekta Parishad also builds upon transformative work on social justice and peace and rehabilitation of dacoits in Chambal Valley in which noted Gandhian workers such as Dr. S.N. Subba Rao known popularly as Bhaijee and P.V. Rajagopal had taken part. Rajagopal was then a young man and he worked with Bhaijee and others to build an Ashram in

Chambal Valley to rehabilitate the Dacoits. Rajagopal then came to Naga-
land and then to Odisha and worked among the fisherfolks of Chilika.

Ekta Parishad has arisen out of Gandhian streams but it tries to grapple
with contemporary challenges of transformations in India and the world.
Dr. Karl-Julius Reubke is a devoted participant in the work of Ekta Parishad as
well as a keen observer. Before taking part in the work of Ekta Parishad,
Reubke has been involved in the work of Rudolf Steiner and his many-sided
works of transformation in education and society. In his book, *Struggles
for Peace and Justice: India, Etkta Parishad and the Globalization of Solidarity*, Reubke (2020) helps us understand the salient aspects of the work
of Ekta Parishad and Rajagopal which is helpful for us to know. Reubke
tells us that Rajagopal once told him that to understand post-Independent
India or India after Gandhi, we need to realize that Gandhi did not have
only one heir that is Nehru. While Nehru was his political heir, Jayaprakash
Narayan was his social heir and Vinoba Bhave was his spiritual heir. They
had their own spheres of works. Nehru followed a trajectory of development
which created many structures of exclusion and to some extent Jayaprakash
Narayan fought against it. In the initial stages, Vinoba Bhave brought a
spiritual dimension to social transformation. In the late 1960s and 1970s,
both Vinoba and Jayaprakash, for example, worked with the *Bagis* (bandits)
of Chambal Valley to surrender. Rajagopal worked with Gandhians like Dr.
Subba Rao who were part of this process of surrender. On April 14–16,
2022, there was 50-year remembrance and celebration of the surrender of
the Bagis in Chambal valley at Mahatma Gandhi Seva Ashram, Jaura, in
which many of the former Bagis had taken part and many Gandhian and
peace-striving activists and scholars had taken part. Rajagopal had led this
calling for creative and critical memory work and dedicating ourselves to
peace, justice, and non-violence (see Giri 2022b).

In reflecting on his work, Rajagopal tells us how he tried to bring Vinoba
and Gandhi together in a spirit of questioning the existing foundations of
inequality and indignity (see Hogan 2006). He tells us:

> Vinoba was not a radical person, and Gandhi was a radical person but
> in a hundred years people have made him non-radical. . . . Marx is radi-
> cal and Gandhi is some kind of old man walking for peace. . . . So if you
> have to be with Gandhi you have to be just peaceful. I say this is a very
> dangerous projection of Gandhi. What I am trying to do is to draw the
> land reform from Vinoba and get Gandhi's non-violence and radicalize
> these two together in a new form that is applicable to today's reality.
> That is what I have been trying to do since 1980. Can I help young
> people to appreciate Gandhi? Not as a museum piece, but as a person
> who is relevant in today's society, and relevant in the lives of million
> who are suffering? Can I bring Vinoba's agenda of "gift" to the concept
> of rights? Vinoba said, "Please give land as a gift." Now I am saying it
> is not a gift, it is the poor people's right. They have a right to have land,

and you have no right to keep the land because you are keeping more than you should. So turning gift into right, turning the non-violence of museum into non-violence of action.

(Rajagopal in Hogan 2006: 61–62)

Rajagopal has been training young people in non-violent struggles for land rights and dignity by bringing Gandhi and Vinoba together. Rajagopal and Ekta Parishad have been mobilizing and organizing the tribal community to ask for their rights—Jal, Jangal, and Jamin. In the process, Ekta Parishad has mobilized women and has created a large number of women activists.

Rajagopal also tried to bring Gandhi and Ambedkar together. In his work on Ekta Parishad, India, and the world, Reubke tells us that the problems of India after Gandhi cannot be solved within the paradigms of children of Gandhi, and it calls for the grandchildren of Gandhi to find creative ways. Ekta Parishad thus brings different groups together. In Ekta Parishad's Jan Satyagraha, Dalit social action groups played an important role. Ekta Parishad fights for the land rights and dignity of the tribals. On the ground, there is an antagonism between the Dalits and tribals. By bringing Dalits, tribals, and other social groups together, Ekta Parishad is trying to create a common ground for fighting for dignity and responsibility.[15]

Much of the work of Ekta Parishad on the ground lies in working with tribals and other landless people in villages to get right to land. This has been facilitated by the Forest Rights Act of 2006 which gives right to the tribals to get right for the cultivated plot of land in the forest for their use.

A Brief History and Vision of Ekta Parishad

Ekta Parishad has been organizing *padayatras*—foot marches—to fight for the rights of the poor and the marginalized. Even before the padayatra of Janadesh in 2007, Ekta Parishad has been organizing *padayatras* in Chhattisgarh, an area of initial and intense work on the part of Ekta Parishad. During one such early *padayatra* in Pandaria in Chhattisgarh, Ekta Parishad stopped the yatra for a few days to protest against the killing of Biju Baiga, a tribal in the village of Dari Para. He was killed by the forest guards as he resisted their move to evict him and his family. As a result of the protest, Biju's wife got a compensation. The forest guards were transferred and the Government also promised to give land rights to the tribals.

In his reflections on Ekta Parishad, Reubke (2020) presents us many important descriptions and insights about the working of Ekta Parishad and here we can listen to some of these. Ekta Parishad organizes many *padayatras* in Chhattisgarh to fight for land rights of tribals. Ekta Parishad also organizes youth camps. It has a centre for meeting and training at Tilda called *Prayog Ashram*. During one such yatra, friends of Ekta Parishad from Europe such as Reubke had taken part in a community meeting. All present were discussing way of finding solutions to the problems that tribals face. Rajagopal was translating the views of the foreigners to the tribals. But Rajagopal did not present all the views expressed by the foreigner participants to the tribals as he felt some of these may be beyond comprehension of the tribals and they may not engage themselves with such suggestions. Reubke here comments: "I now started to ask whether Rajagopal's approach to change was different even in dealing with expert knowledge" (ibid: 318). For Rajagopal, "A discussion in which [people] cannot participate would completely paralyze their activity" (ibid: 322). It seems that Rajagopal is for transformative participation communication and dialogue which can lead to mutually energizing reflections and actions. During one such meeting,

> Once did he interrupt a young man who had started to explain the real meaning of an English word he was proud to know. Such intellectual explanations kill the atmosphere of dialogue. Rajagopal explained his intervention. Words should be used honestly and be understood with good will. They should not be used to shallow superiority of the speaker.
> (ibid: 322)

During walk, Rajagopal also told Reubke and fellow walkers, "Ekta Parishad can never side with any party, even if one party offers an obviously better solution" (ibid: 314).

Rajagopal has nurtured Ekta Parishad as an organization which has around 200,000 members and 400 full-time workers (Reubke 2020: 610). But Ekta Parishad does not have a formal, centralized structure. During his conversations with Reubke, Rajagopal told him that he expects following qualities from those who would like to work with Ekta Parishad:

> The first one is to be able and willing to work for 24 hours without asking for rest or remuneration. The second one is to have an artistic skill; the third to be communicative and the fourth to be cooperative and able to work in a team.

Reubke was puzzled to hear artistic skill as a necessity for working with Ekta Parishad. During his work and walk, slowly he made sense how artistic skill and participation contributes to a new mode of engagement with social reality and people. Art opens our sense of imagination and participation and it helps us to cross many borders. Rajagopal himself began his life's journey as a Kathakali dancer and in a way Ekta Parishad itself is a dance with self, society, culture, and the world. The padayatras and yatras that Ekta Parishad organizes provide an opportunity for all concerned to dance with self and society in creative, transformative, and transpositional ways and not just reproduce our existing positions of privilege or marginality. Art helps to relate to society as a field of the performative where we act to realize our potential including aspirations, not only a field of determination. Artistic sensibility of the participants help all concerned to realize that yatras and societies as the space of the performative is a space for unfoldment of our potentials together. This also calls for a new mode of participation with each other where we strive to feel responsible for and with the other and not just be preoccupied with our own self-interest. It calls for us to undertake suffering for and with the other with a spirit of joy. As Rajagopal outlines in another lecture, this calls for "wanting to suffer more than others."[16] Artistic ability helps one to communicate and here also communicate one's dissatisfaction without breaking away from the group rather contribute to building a collaborative striving for responsibility what Rajagopal calls team work.

Rajagopal uses "street theatre to communicate with Adivasi" (ibid: 347). People associated with Ekta Parishad also give primacy to artistic expressions and here Ekta Kala Manch, the art wing of Ekta Parishad plays an important role. Ekta Parishad has a centre of training and mutual learning at Madurai called CESCI (Center for Experiencing Socio-cultural Interaction) which is in memory of Maja Keone, a friend and participant in Ekta Parishad's dreams and works. In his visit to the CESCI, Reubke took part in a theatre performance by the street children from Madurai about which he writes:

> The presentation of the street children of Madurai at the CESCI Center in 2002 was probably one of the most touching experiences. The transforming strength of performing a traumatic catastrophe in childhood to overcome its destructive impact in late life was convincing without any further explanation. In Pandaria we saw the touching play about the poor Adivasi beaten up by the three evil gangsters: Corruption, Exploitation and Injustice. The Adivasi is then shown to stand up against them together with his family and friends from Ekta Parishad who come to assist him.
>
> (ibid: 347)

Rajagopal, Ekta Parishad, and its participants use art to overcome inherited deprivations and sufferings and create processes of transformative unity. But despite such visions and strivings for collaborative team work, all such efforts face the challenge of conflict, dissension, and rupture. From his own experience, Rajagopal himself realizes itself as a great challenge. Some of the key participants of Rajagopal who were key participants with him in his struggle have left him. Reubke accidentally met one such person in WSF meet in Mumbai in January 2004 as Reubke narrates this: "[I met Rashmi] the strong lady from Chhattisgarh. Chhattisgarh was the only country, where EP had been successful, she said. The success was actually hers but her part was not properly appreciated. She wanted to be her own leader" (ibid: 358). There have been allegations of authoritarianism, but Rajagopal shares with Reubke that, as the nurturer of Ekta Parishad, he has responsibility to protect the initiative and discipline. When workers use Ekta Parishad to further their own ambition, Rajagopal feels that it is best to leave the organization.

On his part, Rajagopal stepped down as president of Ekta Parishad when he turned 65.

Rajagopal grew up in a village in Kerala, but when he came out to the wider world, he was fortunate to interact with many people from around the world with whom he learnt and developed a life time friendship. Rajagopal spent some time at Sevagram and there he helped one of the inmates from Germany who took an interest in Rajagopal and helped him to learn English. Later on, he met many friends from Europe who supported his work and invited him to come to Europe. Rajagopal has been visiting Europe and many countries, and all these create an occasion for mutual learning. In a recent essay on Rajagopal, Jill Carr-Harriss writes about Rajagopal:

> Rajagopal left Kerala at the age of 20 to go to Seva Gram's Agricultural Polytechnic, a post-Nai Talim college, where he stayed from 1967–1970 to study agriculture. He did not return to Kerala until many years later. He settled down in Madhya Pradesh to do social work, and over the next two-and-a-half decades, he rose in the Gandhian fraternity. In 1993 he became the Secretary of one the eminent organizations, Gandhi Peace Foundation in Delhi. As Rajagopal explained it is not easy for someone who was born in south India to gain acceptance in north India.
>
> (Carr-Harriss 2021: 149)

Reubke has accompanied Rajagopal in many of his travels over the years, and in his book, he shares some of the thoughts and experiences from these travels. In one of his lectures, Rajagopal had said that to respond to the challenges of globalization of economy, we need a globalization of solidarity. This requires a transformation at a much more fundamental level, for example at the level of our desire, such as our greed. Rajagopal said: "Gandhi often used the words not only *ahimsa* and *satya* but also *asangraha*. This non-accumulation emphasizes the importance of overcoming greed in our personal life" (ibid: 372). In one of his visits to Germany in the city of Kassel, Rajagopal took part in a youth camp and he encouraged them to play games and try to find out solutions by thinking, acting, and playing out of box. During interaction with them, he articulated four steps of group dynamics that would facilitate self-development and mutual learning: (1) listen properly, (2) look at the world properly, (3) articulate what you think, and (4) act. While speaking about miseries and deprivation of people in India, for example, the tribals with whom Rajagopal works, he asked the young people to reflect upon what might be their parallel common poverty. "Suddenly they all agreed that their common problem was the inability to listen to the elder generation, to communicate with the elder generation" (ibid: 385). Thus, the suggestion from this emerged that "the attempt to change their situation and in consequence to better the world should start from there" (ibid: 385). Reubke reflects upon this mode of interaction with the youth in this youth camp in Germany:

Slowly I realized that Rajagopal's intention was not a measurable success. He offered a chance to start thinking, he did not compel his audience
to it. . . . [H]e invited them to take Gandhi's part in leadership. It was
an invitation for self-development but at the same time to integrate this
development in the development of "my group" became very clear.

(Reubke 2020: 385)

Walking with Rajagopal and spurred by discussions like this, Reubke
started reflecting on the language and condition of poverty. He realized, as
the young children in the camp, that there are many kinds of poverty. There
are not only material poverties but also moral, social, and spiritual. People
in Europe who come out in solidarity with the material poor such as the
tribals of India do not realize that they suffer from social poverty, moral, and
spiritual as well. There is greater degree of social isolation in many European
societies, and it seems that many of them lack the social urge to meet with
others. Reubke comments:

It took me sometime to accept that I also belong to the Europeans with
a deficiency in spiritual culture compared to the Adivasi communities.
I knew that people with social deficiencies are not aware of what they
are missing. I began to realize that spiritual deficiencies are not only
ignored; they are firmly believed to be a sign of higher evolution. People
affected by this mental disease vigorously maintain they are healthy. . . .
Still spiritual poverty is terrible and quite general scourge of our times.

(ibid: 394)

Such realizations are integral aspects of a journey towards our interlinked
consciousness of global responsibility. Our acknowledgement of various
kinds of poverty that we suffer from has the potential to create a common
bond between us in terms of our shared vulnerability which in turn contributes to our shared responsibility or moves towards responsibilization.

The transnational journey of Rajagopal and dialogues of Ekta Parishad
also create a condition for important border-crossing dialogues. In his European journey, Rajagopal meets with crusaders for social justice and human
dignity such as Nicanor Perlas from the Philippines. Perlas has fought for
rights of land and freedom in the Philippines and has pioneered the introduction of large-scale organic and bio-dynamic agriculture. Perlas has been
influenced by Rudolf Steiner and his vision of Three Folding Order and he
brings this vision to his work. He has created a sustainable city in the Philippines. He has written prolifically on several themes on sustainable development, globalization, and evolution of consciousness such as *Associative
Economics: Responding to the Challenge of Elite Globalization* (1997). In
Europe in several places, Rajagopal and Perlas have dialogued which was
potentially a dialogue between streams of Steiner and Gandhi for transformation of humanity. It may be noted here that Ulrich Roess was an activist

with the Anthroposophic Movement of Steiner who also brings Steiner and Gandhi together for cultivating a new cosmopolitanism (Ross 2018). Steiner had spoken about bio-dynamic agriculture where there is little use of chemical pesticide and this has led to several projects in biodynamic cotton farming. Ross had initiated one such biodynamic cotton growing project in Chhattisgarh and Ekta Parishad and some of its associated farmers were involved in this. Reubke had also visited such biodynamic cotton farms. Thus, Reubke and Ross possibly had already opened up bridging paths to Steiner for Rajagopal and his Gandhian ways. Coming back to Rajagopal and Perlas, both of them had taken part in a dialogue at Goetheanum, the spiritual headquarters of Anthroposophy, in 2013, on the theme of "Forming Alliances." In this workshop, Perlas has spoken thus:

> Forming alliances means forming relationships; this presupposes self-awareness. Now the self has two forces: lower and higher self. The lower self has the tendency to cling onto the past. . . . The higher self is capable of creating new social forms so as to take hold of the problem of the world.

In the same symposium, Rajagopal made a presentation on the theme of "Wanting to Suffer More Than Others" in which he said:

> Our responsibility to act swiftly and collectively is great. No individual can work on their own to have any impact against the effects of globalization. Forming alliances and keeping them alive demands sacrifice and willingness to learn from each person.

He presented Jan Satyagraha 2012 in which 50,000 people had marched from Gwalior to Agra for land as an example of such movement of alliance among individuals, groups, and people. For Rajagopal, it was the largest "non-violent campaign for land, water and woodland rights there has ever been" in human history. Rajagopal also said, "In 2011 I covered 80,000 kilometers throughout India to visit all the places of protest against nuclear power, against mines, for free seed—and to show solidarity."

Both Rajagopal and Perlas emphasized on the need to create peoples' power. Both of them challenge us to realize the spiritual dimension of our struggle for a different world. For Perlas, one great spiritual and related political challenge before humanity is how not to be enslaved by technology and to keep our humanity alive and also evolving. Perlas argues that along with striving for justice and dignity, one great challenge before humankind is to confront the technological challenge to humanity itself. There is a technological convergence now of coming together of biotechnology, nanotechnology, and communication technology together which reshape our idea of being human and it may also make the human species extinct. In an interview available in YouTube, Perlas tells us that the whole question is one of technological singularity and its domination of human. It raises the fundamental question of what it means to be human. Unfolding technological convergence raises questions about who we are, where we are, and what is human nature. We need to have collaborative technology which should involve not only convergence of different technologies but also mutual dialogue between inner self and outside forces. We should rely not just on technology but also on inner self-mastery and bring inner change and structural change together.

For Perlas, this technological challenge to humanity and a dominance of machine over man is a far greater challenge than climate change. Like Rajagopal, Perlas lays a lot of emphasis on the youth as a possible harbinger of a new future as they have the natural urge to be creative and not just follow a mechanical model of life. Young people who are growing up young in the new millennium for Perlas have a different orientation to life. According to a study, millennial generation is not interested only in making money but they want to serve the world. They are starting up varieties of incubators of social business to create businesses and social enterprises which would help them not only earn their living but also serve the world.[17] If others in the millennial generation do not find a way of being with the world creatively, they turn egoistic. In countries like the USA, the challenge is to bring this generation and the earlier generation of baby boomers together with the attendant transformation of egoism which would be a great force for transforming the world.

Like Rajagopal, Perlas believes in developing people's power as a way of transforming state power. For him, the challenge for us is to move from a state-centred model of governance to a society-centred model of governance where people inside the political system also strive to live by values that animate civil society. Such people's power can play a role in not only within states but also transnationally. As an example, Perlas draws our attention to the fact that, when WTO began in 1995, it began with a sense of triumphalism that it would run the world. But soon protest emerged beginning with Seattle which changed the dynamics. It may be noted that the WTO protests in Seattle was a defining moment in anti-globalization struggles which inspired subsequent protests in WTO meets. It also inspired movements such as ATTAC and protests in G8 summits. I had taken part in the WTO meeting in Hong Kong in 2005 as a citizen of the world as well as a student of world alternatives. I had taken part in assemblies of protests and alternative discourse there. One of the protest actions involved farmers from Korea jumping into the river in Hong Kong City as a way of protesting against WTO policies. Another show was a fashion show which brought to the audience the plight of cotton farmers in Mali because of subsidy granted to farmers. So, it is the processes like this which Perlas has in mind when he talks about halting the giant of one-sided globalization unleashed by WTO and the work of civil society groups in the world for creating a different possibility of life, wealth, and power.

Rajagopal stresses on developing solidarity with the people, especially the forgotten, in the so-called global development work. In his work, he feels that Northern NGOS and funding organizations do not understand the need for people's struggles for rights. In this context, Rajagopal writes:

> Northern NGOs somehow fail to perceive the spirit of struggle that animates people's standing up for their rights, claiming what is their, right and standing their ground in the face of oppressive state power. They fail to see the larger picture and are merely concerned with statistics that anyway give out only a skewed picture of ground reality.
>
> (Rajagopal 2005: 467)

Northern NGOs and funding organizations emphasize too much on project funding which calls for accountability for every rupee spent but in doing people's work and supporting people's struggle, what is important is trust-based funding. Friends of Ekta Parishad from Europe have supported Rajagopal and Ekta Parishad to work without emphasizing on accounting every penny spent.[18]

Before Jan Satyagraha 2012, there was an international seminar on non-violence at Vishwa Yuva Kendra Delhi in September 2012. This brought together activists, thinkers, and spiritual seekers such as Achan Sulak Sivarkasha from Thailand. It was a rooted global dialogue on non-violent movements for transformations. Before this, Rajaji had undertaken a year-long padyatra across India in which many people and local groups in their communities. It was a journey of *tapa*,[19] *tapasya* of fellow-being, engagement, and understanding. Rajaji shared the following in this international seminar:

> I have come from a long journey all across India. I have seen invisible forms of exclusion about which nobody talks about. Social problems such as children of Devadasis not being able to marry. In spite of our work in the social field, we do not reach out to the last person. I met many others who are invisible. Inspite of forty years of work, I had no knowledge. I realized how miserable the lives of transgender are. They do not have any right. I met a group of people called Siddhi who were brought from Africa 500 years ago. Tea tribes in Assam: they are still working like slaves.

About non-violent social transformation, Rajaji offered the following challenge:

1   How do we graduate from resisting non-violently to using non-violence to create a peaceful society
2   How do people behave non-violently and create a non-violent society
3   Human reality: how do we develop non-violent relations with non-human animals and Nature
4   If we have documentation of non-violent practices in terms of creation of non-violent society, it would help people to graduate from one level to the other.

Rajaji also said:

The basic emotion is anger. In our present-day world, everybody is angry, nobody is happy. How do we use anger and direct it into non-violent action. We need a lot of trainers for diverting anger for positive action.

One can graduate. We act non-violently but we still hate. We act without hating the person in front of us. The exploiter is the victim of a thinking process.

The meaning of education is to develop qualities within oneself, to remove fear from oneself. Organizing the village is the most important challenge in democracy. After the surrender of the Dacoits in Chambal Valley, we started doing this by setting up an Ashram[20]

People engaged in non-violent actions are creating conflicts. How much can I create and handle? This is also a part of training and more networking. Conflicts have to be brought up in order solve it.

Global forces are bent upon land, forest and water for profit. People want it for survival. How can social movements train people to use non-violence to save livelihood resources? If all the resources are grabbed by corporate sectors, the only way left for people is violence.

It is not easy task. I began my work with Subba Raoji twenty years ago. There is no short cut to going to villages, mobilizing against the state. The challenge is to use non-violence at the village level and then scale it up. For Jan Satyagraha 2012 for a people of 100,000 taking part we trained 12,000 young people.

In India, 70% of people are on land. Land is identity, dignity and security. Moving from landlessness to landownership makes a [big] difference.

Social movements are divided ideologically—Gandhian, Ambedkarite, Socialist, Marxian. . . . In Jan Satyagraha, we have tried to bring them together. In a seminar, it is good to fight over ideology. But for the people, the question is how do we survive in a world where forces are out and out to grab my resources. We need to take control of our own egos, ideological egos. It is not just one action—multiple action. During this one year padayatra, we went to struggles—small and big. We organized hundreds of public hearings. We were collecting stories from all the struggles. 2000 organizations are now on a platform. Creating a pressure that non-violent action is not just one action but multiple action. Not only organizations, hundreds and thousands of people have been saving one rupee a day and one handful of rice to take part in Jan Satyagraha.

Sangharsh, *Sambad* and *Rachna*—struggle, dialogue and constructive work—all three have to be involved. In order to survive, we need to have all the components of our struggle. We need to move beyond struggle to dialogue. While continuing the struggle we need to engage in dialogue.

Violence is becoming an arena of employment. Companies are recruiting youth to unleash violence on people. Small arm producers are selling arms all across. The challenge is huge but possibilities are also there.

After Janadesh 2007 we had a National Land Reforms Committee and we received 300 recommendations. Land became an agenda of discussion. The Forest Rights Act got final clearance in 2008.[21] Because of this Act, 12 lakh people got land.

In 2007, people got killed in the accident. 25,000 people instead reacting sat on the mediated on the road.

I tell the Government: Government should have some capacity to deal with non-violent movements. [It is unfortunate and strange that] There is no peace department while there are police and defense departments. There is glorification of violence now and we need to realize that between silence and violence there is active non-violence.

In this pre-Jan Satyagraha seminar, there were activists and thinkers from other parts of the world. They also walked with Ekta Parishad for a few days. Xavier Renoue is a non-violent activist from France who has been involved in many struggles in France and Europe. In his presentation, among other things, Renoue said:

Poverty, including extreme poverty, is coming back to Europe. . . . We are moving from a pacified situation to violent confrontation. State has become lawless and people have become violent.

Renoue also said how Black Block is a tactic used by violent anarchic activists. It is an offensive tactic in which they form a dense bloc. But in its place, Renoue and fellow non-violent resisters have chosen civil disobedience. Resonating with his French secular tradition of criticism and emancipation, Renoue said that there is no such thing as spiritual culture of non-violence. We are mostly atheists. We do role playing game. We show them how powerful non-violent technique is. But in the process, people move from a tactical perspective to a philosophical perspective.

Coming right after Renoue, Sulak Sivaraksha from Thailand said:

Activists and secular intellectuals think that all problems have been created by the system without realizing how all these problems also exist within themselves. Transforming society requires personal and spiritual change first. Those who want to change society must understand the spiritual dimension of change. Social change and spiritual consideration cannot be separated. We cannot overcome the limits of individual self in a hermetically sealed environment. . . . Buddhist practice begins with mindful breathing. With awareness, we cannot solve problems alone. We need friends. . . . Breathing is the most important element of my life. I learn here the breath of peace. I smile as I inhale, as I exhale I also smile. Vipasana meditation create samata bhavana—feeling of equanimity. It helps us avoid taking ourselves too seriously. We seek peace and justice on a real understanding of ourselves and the world. We are not governed by greed, hatred and ignorance.

Speaking about the spiritual dimension of non-violence, Heather, a participant from Canada, said:

> I started my spiritual journey with *L'Arche*. [It is an initiative to nurture the differently able and autistic people in Canada and around the world drawing inspiration from the work of the noted philosophe activist thinker Jean Vanier. . . . But I realized that this kind of spirituality is unable to deal with structural violence. I went to South Africa to study with Gustavo Guttiriez, the founder of liberation theology. I came to see the world in structural patterns of opposition. . . . What happens when we have democracy representing corporate interest and not people? This is when critical Marxian analysis helps.
>
> > Turning to Gandhi, non-violence is not a technique but a way of life. What I like about Gandhi is that it helps us to diffuse ideological boundaries. As women, we need to have sophisticated understanding of boundaries. . . . And for understanding the question of land, we can build on both Gandhian and Marxian analysis.

In her presentation, Heather challenged us to realize the significance of the aesthetic dimension in social justice movements:

> There are so many social justice movements but there is reduction of beauty in the world which diminishes our capacity to live. A culture of peace has a spiritual depth. The spiritual dimension requires all of us to cultivate interiority. It opens our awareness. They awaken us to levels of reality that I did not see before.

Legborsi was a participant in this dialogue from the Ogoni land and is associated with the struggle for rights and justice of the Ogoni people initiated by the legendary activist, thinker, and martyr Ken-Saro Wiwa (Wiwa 1995). Wiwa had struggled for environmental justice and democracy in Ogoni land and Nigeria and had fought against Shell. He was falsely accused of murder and executed by the military dictatorship of Nigeria in 1995. Legborsi continues the dreams and struggles of Wiwa and other martyrs of the land.

Kaima was a politician of the Green Party from France. Kaima had done a lot of work to create support for Ekta Parishad in France. In the seminar, she said:

> I was born in a poor family in France. Each day when you wake up, it is a struggle. In Europe, all over the world, people forget this. Our mission is to promote a new model of the common ground.

After the seminar, there was a public meeting at India Habitat Center in which some of the participants of the seminar as well spoke. Mr. Jairam Ramesh, the then Minister of Government of India and in charge of rural development, also spoke. He told the audience the dialogue he had with Rajaji and the activists of Ekta Parishad over the last year in various places including at Prayog Ashram, Tilda. He spoke with a sense of humour:

> Rajaji was a Kathakali dancer. Kathakali dancers dance slowly and in the last year he made me dance in so many places.

With all these, it was all placed to start the much-awaited Jan Satyagraha from Gwalior on October 2, 2012. There was a big inaugural meeting which was addressed by Rajaji, Swami Agnivesh, and by two Central Ministers— Jyotiraditya Sindhia and Jairam Ramesh. Jairam Ramesh, as the main point of Government's negotiations with the Satyagraha, had assured that in the big rally, he would make the announcement that Government had accepted the demands of Jan Satyagraha. So, the marches can go home. But in the rally, he made an out of turn speech much to the dismay of the organizers. He said that Government is trying to do its best and the marchers should rely on the Government and go home. Having come thus far and from all over the world, marchers should see the historic city of Gwalior and come home. But towards the end of the inaugural session, Rajaji responded in a creative way which reflected deep spiritual attunement and political wisdom. He said that this is the Satyagraha of the people. Let people decide what they want to do. He asked the assembly whether they would like to go back home or continue the *satyagraha* march. The assembly unanimously thundered: *Adharoti Khayenge, Phirvi Ham Chalenge* [We would eat half breads but we would still continue our Journey]. It was a great challenge for Rajaji, Ekta Parishad, and other participating organizations as on the assurance of Mr. Ramesh, Ekta Parishad had not organized resources to undertake the march. To help run 50,000 people on the road, make run supportive vehicles, and arrange food—Ekta Parishad had not managed money for this. But Ekta Parishad and friends were not disheartened. Rajaji and key leaders of Ekta Parishad including from Ekta Europe met in a tent in the evening of October 2 in the Gwalior Ground and discussed ways of meeting this challenge of resource mobilization. Some of the resourceful friends such as an industrialist from Mumbai and a devoted walker and worker with Ekta offered resources to take care of the food requirement. Some friends from Ekta Europe offered to take care of the diesel for the journey.

The next morning the march began. We all around 50,000 people started our journey. Rajaji and the core activists of Ekta Parishad were on the front followed by activists and other marchers coming from different regions and states of India. Different state units formed their own column. There was a truck attached to a state unit which carried the luggage of the marchers. It also carried the food ration and the kitchen utensils for the unit. The march would stop for some time when a nearby town or village would like to come and welcome the march. Usually leaders of the local community would come, welcome the march, offer their support, and garland Rajaji and other leaders. Rajaji and other leaders would also speak about the purpose of the march and the larger struggle for land, human dignity, and liberation lying with and ahead. Marchers would also sing songs and offer creative cultural performance such as dance and music. It was a walking mobilization.

In one afternoon, a group of local engineering students came to greet the marchers. Rajaji was resting by the road side. He spent gracious moments together with these young people discussing in a friendly and involved way values of life. What is the purpose of education? How can one use education for helping others. In one evening, near a roadside restaurant, in front of some *charpais*, there was a meeting with visiting local leaders and *sarpanches*. He again discussed with them the values of politics and political life and the higher purpose of life.

There were varieties of participants from various walks of life. During the march, I spoke with them. In one afternoon, I was sleeping in the field with an elderly person from Bihar. He was accompanied by some of the villagers from his village. He himself was not feeling that well. It has been tiring for him walking. He was under the impression that joining this march he would also get to see more tourist places. The heat and illness added to his disillusionment. He was asking himself and a fellow being why he was here and when he would come back home. In fact, some people were leaving the march, as they felt that the march was moving on out of exhaustion, frustration, and lack of purpose. In fact, during the marches, Rajagopal and the top leadership of Ekta Parishad used to have a reflection and evaluation meeting. In one such meeting among closed confidantes, Rajagopal had raised his concern as to why people are leaving from the middle of the march.

Because of the sudden nature of continuing the march, there were a lot of contingencies on the road. One challenge was the inadequate health care for the marchers. As can be imagined for a marching humanity of around 50,000 or more, there was not proper health care available on the road. There were not enough volunteer doctors on the road to help the suffering marchers. Rajaji acknowledged this challenge and counselled patience and care from and for all concerned.

In the morning, the walkers have a cup of tea and start the march. During mid-day around 12:00 or 1:00 PM, the march would stop on the road. Walkers would find a shelter near the road. The accompanying truck carrying ration and food would have arrived earlier and food for the regional marching group would have been prepared by a set of volunteers. Then the marchers would take a bath, freshen themselves, and come to eat. After lunch, they would rest. The afternoon and evening is a time of mutuality including mutual creativity. Fellow marchers would sleep side by side each other and tell the stories of their lives. In the evening, they sing songs, do dramas. They would also listen to and make consciousness raising speeches.

In the evening marchers also would visit other regional groups. In different regional groups, there was performance in respective mother languages. It was a great occasion for the marchers to not only hear but also experience the multi-versal and multiple live cultural streams of India. Jan Satyagraha thus became a walking school and a festival of life.

During our walk together during Jan Satyagraha, I had a discussion with Jill Behen, wife of Rajaji, and a leader of Ekta Parishad, about the challenge of responsibility in movements of justice. I suggested that responsibility is not emphasized enough in people's struggles. Jill Behen made an insightful comment:

> I began with giving charity but then I discovered strength with people. When you find strength with people, you see them as responsible. Then they can take action. Once people consider themselves only powerful and [others not] then they feel that only the powerful can take action.

On the first night of the march, some of us marchers were resting in a local police station. Aditya Bhai, Mr. Aditya Patnaik, a Gandhian creative worker from Odisha, was a fellow participant both in the international seminar and in the march. In fact, on the first day, he was walking with a spirit of radiant and divine solidarity with Rajaji and Radha Behen, another radiant soul and then President of Gandhi Peace Foundation. At 3:30 AM in the morning on the second day of our march on October 4, 2012, I received an email from Aditya Bhai that he is going with a delegation from Gandhi Peace Foundation to meet the Prime Minister to consider the situation urgently. Aditya Bhai asked me to draft an appropriate letter to the PM. I had my minicomputer with me and drafted a letter and sent it to him at 4:08 AM in the morning. As is known, Jan Satyagraha ended on March 10 at Agra with a signing of the Agreement between the Satyagraha and the Government of India. May be this letter of a wandering being at four in the morning had a role in it.

There was a signing of agreement at Agra. But after the Agra meet, it seems nothing much happened. But Ekta Parishad has organized another big yatra from Delhi to Geneva on October 2, 2019. It is called Jai Jagat Yatra. The inaugural meet was held at Viswa Yuvaka Kendra in New Delhi on October 2, 2019 with some from the last 2012 pre-Jan Satyagraha being present such as Achan Sulak Sivaraksha from Thailand. The walkers travelled in vehicle from Delhi to Jora in Chambal valley in Madhya Pradesh where Rajagopal had started his early work with the local dacoits and their surrender to the police and non-violent change.[22] Walkers stayed in the Ashram established by Rajagopal, Subba Rao, and others then. They started their footwalk from Mahatma Gandhi Ashram, Jaura, to Sevagram, Wardha, on October 11, 2019, the birth day of Jaya Prakash Narayan, and reached Sevagram by January 30, 2020. Jai Jagat Yatra had a meeting on this day of martyrdom of Gandhi in Sevagram. After this, the walkers formed into groups—some went to United Arab Emirates, some to Bangladesh, some to Pakistan, and some to Iran. They all flew to Armenia to continue their walk to Geneva. But because of COVID-19, the walkers stopped their walk in Armenia in March

2021 and got back to their respective countries luckily before severe lock down started happening in their countries including in India.

In a recent meeting in April 2021 at Jai Jagat Center in Puducherry, I was speaking with Rajagopal about this walk. Rajagopal told me that this walk was motivated by Gandhi and Vinoba's spirit of well-being of our entire world. It also resonated with some of the goals of United Nations Sustainable Development Goals such as reducing poverty and establishing peace. It wanted to create a global foot walk campaign for peace and justice. Rajagopal had gone to Iran before coming to Armenia from Sevagram and he told me how travelling through Iran was a great experience for him in cross-cultural experience of India and Iran which also reminded him of age-old Indo-Persian connections. He and some of his co-walkers travelled from cities to cities and met common people. In Armenia, one group of co-walkers were already walking and they collaborated with the peace workers of Armenia. Rajagopal and his co-walkers from Iran and other co-walkers from Bangladesh, UAE, and Pakistan joined them.

Rajagopal is an inspiration to many of his co-walkers and co-workers. His wife and researcher Jill Carr-Harriss writes:

> When he started talking, his smiling eyes did not reflect the many challenges he faced in his life; rather he brought Gandhi's values to life right in front of me. He spoke about how mass mobilization uses nonviolent techniques and how they had mobilized thousands of disenfranchised people. . . . As I listened to the unbelievable description of thousands of people being mobilized, I wondered how a mild-mannered and such a respectful person could lead such large numbers of people; and at the same time, confront hardened politicians that only knew rough language and political expediency. Rajagopal made the point that strength does not come through provocative language.
>
> (Carr-Harris 2021: 149–150)

In a recent discussion in March 2022 at CESCI, Madurai, Jill Behen told me that Jai Jagat emerged out of a realization that grass roots struggles for land for the marginalized in India needs to be linked to global efforts in communication of such struggles to global agencies and building of global solidarity among like-minded individuals, groups, and movements.

Ajit is a young volunteer of Jai Jagat campaign whom I met at Jai Jagat Center in Puducherry. Ajit had studied architecture and comes from a Gandhian family in Kerala where his mother had instilled in him a love for Gandhi and his art of non-violence. Ajit had studied architecture and he wanted to practice non-violent architecture which is people friendly as well as eco-friendly. Ajit joined the walk as a new walker. Among the 50 people who had joined the Jai Jagat walk from different countries, there were 15 people who were new comers and Ajit was one of them. For him, it was a great social and spiritual experience for him to join the walk. During the walk, he felt the moral strength of Rajagopal, his fellow walkers, and Ekta Parishad. He also met a woman from France during the walk who nurtured him like a mother. After the walk, Ajit has joined the Jai Jagat Center in Pondicherry as a co-ordinator. He is turning also towards spirituality and during our conversation he told me how he prays God to make him an instrument of peace and do something for society. It is the Jai Jagat yatra which has enkindled in him both love for society and spirituality and during our conversation he told me, "I want to find the space between Ramana Maharishi and Rajagopal, Sri Aurobindo and Gandhi."

Because of COVID-19 outbreak, Jai Jagat yatra could not continue but still in September 2020 some marchers from Europe could go to Geneva as planned. Carr-Harriss writes about this:

> In September 2020, the European marchers who had been with us on the global march engineered a walk from France to Geneva and in spite of COVID restrictions, one hundred and forty people from seven locations entered into Geneva on the very day that our Jai Jagat march was supposed to arrive. The group went to the UN Headquarters to deliver the message of the Jai Jagat and had a meal together before dispersing.
> (Carr-Harriss 2021)

Yatra has been an important mode of engagement with Ekta Parishad, but there are some critical notes on this Yatra mode of doing politics and fighting for social change: is bringing people on the road is effective in achieving this objective? Here I present two critical perspectives. A participant in our international conference asked me after a few months of Jan Satyagraha 2012: Does Ekta Parishad have the strength to mobilize people and fight against the Government if Government goes back on its promises? Another observer of the scene, a professor of Economics from JNU, commented during a meeting and conversation: "If Rajagopal believes that you can solve the land problem by talking to Jairam Ramesh, then . . ."

Jairam Ramesh had a role to play in formulating the new land acquisitions act of 2013 which called for wider consultation and consent of the affected people before acquisition of their land.[23] Mr. Narendra Modi after getting elected as Prime Minister tried to change it. He proposed to make it easier to acquire land. There was a country-wide protest against it. Ekta Parishad also played an important role in this protest. The ruling BJP Government also did not have the required majority in the House to pass the proposed new amendment to the Act. Finally, Mr. Modi left it to State Governments to enact their own laws. It was a victory of some sort.[24] But at the same time, the legislation of many other State Governments such as Odisha which wants to make encroachment of Government land a criminal offense makes the poor who do not have any option but to build their huts on Government land deeply vulnerable.

The Work of Ekta Parishad

During our field work, we visited several sites of the work of Ekta Parishad. We began this journey of understanding and realization with Prayog Ashram, Tilda, Madhya Pradesh, in September 2012. It was before Jan Satyagraha 2012. There were many people from the local community. There was discussion on how to take part in Jan Satyagraha in a spirit of peace, service, discipline, and dedication. We also visited two villages where Ekta Parishad was active. In these villages, tribal people were engaged in their efforts to get *patta* for their forest land as a result of the 2006 Forest Rights Act.

In fact, working with the Forest Rights Act to mobilize villagers to get land was an important mode of mobilization for Ekta Parishad. This reflects the motto of Rajagopal: "We cannot let communities vanish like ghosts" (in Helena Drakakis 2005: 18). We visited the work of Ekta Parishad in a village in Rajgangpur Sundergarh district of Odisha. This is a tribal village. The work of Ekta Parishad mainly happens with the work of a sister here who is employed by Ekta Praishad as an animator. She works with local villagers to get the *patta* of the forest land that they have been cultivating. Kahnu Bhai is a leader of Ekta Parishad in this area. During our visit, we had a meeting in the village hall with some villagers about the values of tribal people and the work of Ekta Parishad. Kahnu Bhai said that tribal people still reflect a way of living harmoniously and in the name of development we should not destroy this. From Rajgangpur, I had come to Patna in December 2013 and visited the office of Ekta Parishad there. Ekta Parishad was trying its best to mobilize in a difficult climate of lack of political will in Bihar to implement land reforms. But the workers working there were animated by a sense of purpose. They were thinking of varieties of ways of keeping the land issue alive and active.

From Patna I had come to Bodhgaya where Ekta Parishad was active. I visited a village along with local activists of Ekta Parishad where some landless families have occupied a vacant Government land and had built their houses.

On February 16, 2014, I had visited *Naya Sabera*, a voluntary organization in Hazaribagh which collaborates with Ekta Parishad. Ekta Parishad along with *Naya Sabera* here works on land, forest, and livelihood. It works with people on land improvement and land mobilization. It works from rights-based to constructive work. *Naya Sabera* works with SRI (system of rice intensification) system of rice production which increases yield in rice cultivation. It makes seed banks and it also works with farmers to have a soil health card. It has created 700 to 800 SHGs (self-help groups). It has also helped 220 people to receive *patta* for their lands. Along with liberation from hunger of the villagers, it is also working towards local self-governance. It organizes youth camps where campers both work and study bringing plough work and intellectual work together. It also tries to regenerate fallen forests. Ram Swarup Bhai is an activist and worker of Ekta Parishad in this area.

Ekta Parishad works with tribals and many of these areas Maoist forces have also been working. Ekta Parishad began working in Chhattisgarh. It also began working in Bastar region which has now turned into a war zone between State and Maoist forces. In the middle of it, State supported vigilante groups such as Salwa Judum which unleashed reign of terror on innocent tribals and forced them to leave their villages and live in camps. As a result of a PIL (public interest litigation) filed in the Supreme Court by Professors Nadini Sundar and Ramachandra Guha, the state was forced to disband Salwa Judum but this happened only in name. According to a recent report, the whole of Bastar has been turned into a jail where State and Judiciary have come together in the name of fighting the Maoists.[25] Maoist forces are also killing innocent people in land mining explosions as well as in the name of getting rid of police sympathizers. In this theatre of violence, Ekta Parishad has been driven out. But in the beginning in the 1990s, it was Ekta Parishad which was one of the few organizations fighting for tribals in non-violent ways. It brought tribals out on the road, *sadak par ana*, as a way of self and community mobilization and fighting against the State. It, for example, fought for fair prices for minor forest products such as Tendu leaves. But according to Komal Kankar, a sensitive journalist from the region, with the change of Government in state of Chhattisgarh, the Government ruled by Bharatiya Janata Party did not give space for the workers of Ekta Parishad to work in Bastar.[26] It foisted false cases against key leaders of Ekta Parishad in the region which made them impossible to function as they became embroiled in court cases. Ekta Parishad was annihilated in Bastar by the political establishment using state and judiciary machinery which then only encouraged Maoist forces to grow. But the issues that Ekta Parishad had raised in the beginning are now being taken by the Maoists in a different way. According to Kankar, while all the major political parties such as BJP and Congress and their Governments are bent upon plundering the land and resources of the tribals, only Ekta Parishad and the Maoists think about the issues of the tribals—their rights to their land and resources—and fight for these.[27]

Professor Manoranjan Mohanty is a deep thinker and activist of India and the world who has been involved with many grassroots movements of change. Mohanty had taken part in the concluding rally of Janadesh 2007 meet in Delhi. I had a discussion with him about his perception of the significance of Ekta Parishad in the struggle for land and social transformations.[28] He said that Ekta Parishad works as a social movement and voluntary organization and the marches have created pressure on the Government to initiate some changes. However miniscule it may seem, this does play an important role in the transformative dynamics. But Ekta Parishad is not like MST of Brazil. It would not think of occupying land. It is also not like the *Chasi Mulia Sangha* of Narayanpatna, Odisha, where Nachika Linga can occupy land.[29] But this land occupation movement which was autonomous was subjected to unprecedented state repression which let this to be taken over by the Maoists. While Maoists movements are challenging state occupation of people's lands, the violence from both sides creates unprecedented misery. Ekta Parishad's non-violent methods and struggles have an important role to play in changing such conditions of violence and continuing the struggle for land justice and dignity.

In my most recent meeting with Rajagopal, Rajagopal reiterated his and Ekta Parishad's commitment to using active non-violence as a tool for social change. Rajagopal is now focusing on training young people for this. After the COVID-19 lockdown since March 2020, Ekta Parishad has been focusing on providing support to the migrant workers in India coming back home. As may be noted, the total lockdown announced by Government of India at a short notice of four hours on March 24, 2020 by Prime Minister Narendra Modi on behalf of Government of India created untold misery, suffering and death for many migrant workers some of whom died walking hundreds and thousands of kilometres to reach their homes and villages (see Mander 2020). With the help of its local units, Ekta Parishad created support for the returning migrants. It created labour banks where migrants can get grains for their use and in return deposit their labour in the Labor Bank or Shram Bank. When they get work, they return the cost of the grain borrowed from the Bank.[30] On the onset of second wave of COVID-19 in India from April 2021 which is also affecting many villages across the country, Ekta Parishad is organizing relief and medical support for the needy.

Transnational Work of Ekta Parishad and Globalization of Solidarity

Ekta Parishad has supporters from many countries in Europe. In fact, there is an Ekta Europe which provides support to Ekta Parishad's works. It holds regular consultation meetings and arranges travels and lectures for Rajaji.

Padayatra—foot walk and foot march—is a method of Ekta Parishad. Some friends of Ekta Parishad have also been organizing padayatra in France. Ekta Parishad's *padayatra*, foot work, has been put into practice in other places as well. There is not only a transnational participation in Ekta Parishad's padayatra but Ekta Parishad's padayatra has also inspired friends in other cultures and countries to arouse consciousness of people about pressing local and global issues. For example, in July 2010, Dialogue of Humanity, an initiative to create dialogues among people, was organizing its annual meeting in Lyons (more about Dialogue of Humanity later). Coinciding with this meet which brought many people from around the world, Manu, a local social worker from the neighbouring city of Grenoble, France, who had visited Rajagopal and Ekta Parishad earlier, organized a Dignity Walk. I had a walk with Manu in his home town of Grenoble. He told me that he had visited Rajagopal in India the year before and by speaking with him and by being with him he could feel the significance of walk for mobilizing interested people and generating consciousness. During their time together, Rajagopal helped him realize that there was no point in just taking a walk. It would be helpful and energizing to organize a walk around critical issues such as poverty: "We should take a walk to fight for poverty and to arouse consciousness about it." Manu has been interested in poverty as he himself grew up in a poor neighbourhood in France staying in poverty housing. Years ago, Manu had also stayed in a slum in Mumbai. For Manu, poverty in France is invisible. It is inside the walls. There is no job for many youth in poor neighbourhoods. There are also problems of single parent household, drug, and violence. Manu works with social centres of Grenoble to help marginalized and excluded youth. Most of the social centres work on personal development: yoga, dance, and cooking. But as a social worker, he feels that social centres should also work on emancipation, rights, and duties. This is more difficult than teaching cooking lessons. Such a broad understanding of his work as a social worker possibly has emerged from his visit to works of Ekta Parishad in India and his subsequent continued interest in its works.

In this walk, friends of Ekta Parishad and people interested in issues of poverty locally and globally joined in. Sylvie was one of the participants in this dignity walk. During our meeting in her home in Chambery, she described her experience of being with this walk: "We walked in city Chambery and across the road to Lyons. We slept in the tent. Some people had never slept outside while others are used to sleeping with the stars. We all slept together on the road."

Along with organizing *yatras* and struggles for land rights, Ekta Parishad also mobilizes support for people during moments of calamity such as Tsunami in 2004. Here again Ekta Parishad's approach was not one of charity but empowering people to take care of themselves. Ekta Parishad also organized children's camps.

Landless Workers' Movement (MST, in Portuguese Movimento Do Trabalhadores Sem Terra), Brazil

Landless Worker's movement (from now onwards MST) in Brazil has been an important land rights movement in the world. It has important parallels with Ekta Parishad in India. MST started in 1984. It has an estimated informal membership of 1.5 million people across 23 of Brazil's 26 states. It started occupying land for the landless families. The first land occupation took place in October 1985. The Supreme Court of Brazil gave a verdict that occupation of unused land for landless families is not illegal, thus paving the way for its works as an occupying movement. MST creates agricultural settlements for the landless people and helps them also with agricultural implements and loans from the Government. MST urges farmers to have an ecological method of farming. In some places, it also provides them education and health care. For example, a decade ago, MST had 1,000 first- to fourth-grade schools and 100 fifth- to eighth-grade schools. MST tries to build a community not only in the locales where settlements exist but also for the whole movement which MST researcher Wendy Wolford (2010) calls an "imagined community." But belonging to this community happens not only out of love but also out of compulsion. MST draws its inspiration from Marxism and Leninism and it works with tight control over the lives of its members which has led some critics of it such as Zander Navarro of Porto Alegre to talk about the dictatorial dimension in the work of MST. For example, the landless families who are settled with the help of MST are expected to help those who are in waiting and take part in the occupation of land for them. As Wolford argues, "Landed members are often asked to provide food for people living in occupation camps" (Wolford 2003: 7). They are also expected to show loyalty to MST. As Wolford tells us, "During a 1997 demonstration, two supposed MST members were expelled from the movement because they were believed to be spies" (ibid).

The struggle for land and dignity and a fuller humanity in MST has been accompanied by much violence and bloodshed. On April 17, 1996, 19 workers of MST were massacred while doing peaceful protest in Para. This massacre is being remembered by MST and Via Campesina as an International Day of Peasant Struggles. This preceded another massacre at Corumba in 1995. In 1997 alone, confrontation between police and MST workers led to two dozen deaths. As Landertinger writes, "Indeed, in the period from 1998 until 2001, 1517 rural workers lost their lives due to attack from military police or landowners' private militias" (Landertinger 2008: 15). It might be noted here that four activists of Ekta Parishad also have been killed.

MST faced different threats and opportunities in recent political history. When Fernando Henrique Cardoso was President of Brazil (January 1, 1995 to December 31, 2002), it faced a lot more obstacle. It also faced challenges from the World Bank whose plan was to privatize land, and landless people could buy land from willing parties and for this they could get loan from the Government. But this was unacceptable for MST and the struggling landless masses as the interest rate was as high as 18% with the grace period of only three years, and no loans available for seeds or supplies (Landertinger 2008: 15). But when Lula (Luiz Inacio Lulal da Silva) became President of Brazil (January 1, 2003 to January 1, 2011) who came from a working-class background, he initially supported the movement. Under Lula Government, a commission was created to elaborate a plan for agrarian reform. Lula made Plinio da Roda Sampaio, an MST member, a member of this commission. But later on, MST protested against many of the moves of Lula Government when it felt that Lula is enacting policies contrary to the interests of the workers and landless farmers. But in the present political crisis in Brazil (in June 2016) when President Dilma Rousseff has been impeached by the Senate for financial irregularity, MST is offering support to the left-leaning political forces of Lula, Rousseff and the Worker's Party. MST and many left-wing thinkers and movements look at impeachment as a right-wing conspiracy to reverse the progressive and emancipatory politics of the left-leaning political forces of Lula and his successor and take power by dubious means. Before and during the impeachment process, MST protested against such a move and it brought its people to non-violent strike at the protest ground in Brasilia. In the meantime, Lula was imprisoned in a corruption case which was a political conspiracy to destroy him and his party and came out of prison from Curitiba on November 8, 2019. He has just won the Presidential election conducted on October 30, 2022 and is striving to heal the deep divisions and antagonism in contemporary Brazilian society and polity.

We saw that in its fight for justice and responsibility, Ekta Parishad has been following a path of building alliance which is not just alliance of politics but a mode of alliance of life. With Ekta Parishad, alliance building has a spiritual dimension of sacrificing one's ego and giving space to other fellow co-travellers and fighters. MST also strives to nurture its own path of spiritual unfoldment, as it draws not only on classical Marxism and Leninism but also on mysticism of Liberation Theology. As may be noted, base communities as part of a liberation theology sprang up in Brazil and other South American countries as communities of striving for justice, peace, and solidarity which brought Christian struggle for justice and Marxist struggle for emancipation together. With liberation theologians such as Leonardo Boff who also works in Brazil, it also took an ecological turn. MST's practice of agro-ecology and its drawing on the mystical streams of liberation theology bring a dimension of practical mysticism to its work which can also be looked at as an aspect of practical spirituality which strives to bring land and soul, food and freedom, soil and soul, and foot and sky together.[31] As Wolford writes:

> One of the main methods for actively presenting the frames underlying MST's community is through what the movement calls "mysticism".... Mysticism is a legacy of Liberation Theology, which relied on charismatic leaders who were able to re-engage people in the practices of the Church. MST activists build on this combination of worldliness and idealism by creatively using songs, theatre and chants to help form new ideas and mould behaviour. Symbols of the struggle for land that characterize mysticism include dramatic representations of a joyful harvest where people work together to bring in the crops, and visits from past resistance leaders. In 1998, one State-wide meeting in Santa Catarina opened with several children walking single file through the audience carrying the tools and fruits of working the land—a machete, a handful of beans, a large squash. These were all laid at the front on a large outline of Brazil, signifying the construction of a better nation through the practices and values of MST's new community. The dominant messages of mysticism are humility, honesty, conviction, perseverance, sacrifice, gratitude, responsibility and discipline. According to an MST publication, mysticism "reduces the distance between the present and the future, helping us to anticipate the

good things that are coming" (Jornal Sem Terra 102: 3). Movement lead-ers encourage activists to use mysticism to tie settlers more firmly into the movement: "the more that the masses attach themselves to their symbols, leaders and the organization, the more they fight, the more they mobi-lize and the more they organize themselves" (Jornal Sem Terra 97: 3). Mysticism is always present at MST meetings, assemblies and public demonstrations.

(Wolford 2010: 11)[32]

Coming back to the spirit of alliance, like Ekta Parishad, MST also carries out much of its struggle with a spirit of alliance. As Landertinger writes, "In only 9 out of 19 protests between September 2007 and March 2008 did the MST act alone. More than half of the movement's activities involved differ-ent allies" (2008: 13).

I had visited the city of Porto Alegre in December 2002 and had vis-ited a settlement where MST was active which was an hour drive from it. There people had built houses and created agricultural fields in the occupied locality.

Via Campesina

Via Campesina has been an important transnational movement for global justice relating to food and agriculture. It was founded in Mons, Belgium, by a group of farmers' representatives in 1993. Via Campesina has become an international umbrella organization bringing together many groups fight-ing for farmer's dignity and creative living. It defends "small-scale sustain-able agriculture as a way to promote social justice and dignity." As Faustina Torrez writes, "*La Via Campesina* sees agrarian reform as a concern for all humanity and searches for new ways of understanding agriculture based on peasant perspectives" (Torrez 2011).

All these initiatives explored in this study are new initiatives in global justice. Movements such as ATTAC and Ekta Parishad also contain seeds of global responsibility going beyond conventional discourses of rights and justice.

Notes

1 https://en.wikipedia.org/wiki/File:ATTAC_poster_Ameugny_200409.jpg
2 Gandhian view is described in terms of concentric circles with man as the cen-tral point. In Gandhi's views, there need to be "ever-widening, never ascending circles with the individual as the unit." Life will not be like a pyramid but will be a concentric circle where each will depend upon the other. This reality of mutual interdependence makes these circles inter-penetrative and transmutational. Thus Gandhi's idea of concentric circles can also be accompanied by a picture and reality of inter-penetrating circles. Such visions and practices of inter-penetrating and concentric circles are important for realization of responsibility in self, cul-ture, society, nation, and humanity.

3 Democracy is about representation but the key question is how can a representative represent the represented? Ankersmit (1996) raises this question and argues
that we need aesthetic representation and not just mimetic representation in politics and democracy. This aesthetic representation needs to be linked to creative
ways of knowing and being to what I have called ontological epistemology of
participation. It needs to cultivate appropriate self-formation and knowledge of
the other, especially those represented. But even aesthetic representation need to
realize the limits of representation itself. It needs to be opened to the dimension
of emptying, self-emptying, and mutual emptying in self, society, and politics.

4 The Zapatistas are the guerrilla in the Chiapas region of Mexico. They fought
against the Mexican Government in 1995 but after this they left their armed
rebellion and did not fight for seizure of political power. They rather fought for
local autonomy and creative development of the indigenous people in particular
and people of Mexico in general.

5 The vision and practice of confrontative dialogue resonate with my idea of
walking, working, and meditation with compassion and confrontation together
which gives rise to compassionate confrontation or confrontational compassion.
Responsibility, as well as justice and rights, involves such works and meditations
of compassionate confrontation or confrontational compassion. I have begun
exploring about compassion and confrontation in my essay, "Compassion and
Confrontation" in my book *Knowledge and Human Liberation: Towards Planetary Realizations* (see Giri 2013).

6 Solidarity is an important challenge before self and society but it is a process.
There is an integral link between solidarity and responsibility as movements of
rights, justice, and responsibility have a responsibility to nurture solidarity. This
solidarity is neither mechanical nor organic in the way Durkheim and sociologists have talked about it. It is an emergent process which is facilitated by the
work of charity in us and what we have called compassion in this work. Vattimo
(2011: 139–140) challenges us to realize the significance of charity, as he writes:

> At the horizon line of the near future toward which we gaze, pragmatically
> assessing the utility of truth, there lies a more distant future that we can
> never really forget. Rorty alludes to this with the term solidarity, which I
> propose to read directly in the sense of charity, and not just as the means
> of achieving consensus but as an end in itself. Christian dogma teaches that
> *Deus Caritas est*, charity is God himself. From a Hegelian viewpoint, we
> may take the horizon to be that absolute spirit which never allows itself to
> be entirely set aside but becomes the final horizon of history that legitimates
> all our near-term choices.

Charity is also called for in our construction of others. In his reflection on
Christianity, Vattimo (2002) brings creatively an approach of deconstruction to
Christian spiritual quest by telling us that Christianity is a religion not of asser-
tion and violence but of self-emptying and non-violence. Vattimo suggests that
Christianity is unique in this. During our conversation in his office in Turin in
December 2002, I asked him how he can be sure that Christianity is unique in
this journey. There were many initiatives in non-violent religious quest before
Christianity such as Jainism and Buddhism. There are also streams of non-violent
quest in primal religions and spiritualities of the world which Christianity dubs
as violent and animistic. During our conversation, Vattimo felt silent before this
question and he did not repeat himself. Such silence can remind us of the spiritual quest of a seeker such as Adi Shankara from India for whom meditation is
a work of silence and it is not a logic of repetition. It is not a repetition of ritual.
This conversation for me was a deep spiritual experience experiencing profound
humility of this philosopher of weak thought and weak ontology.

In his work, *A Farewell to Truth,* Vattimo (2011) relates the process of building solidarity to what he calls ontology of actuality. This is also an ontology of actualization where actualization is related to meditative verbs of pluralization.

7 Polish sociologist Peter Stompka talks about the process of social becoming.

8 Jean-Luc Nancy tells us that contemporary processes of globalization with the dominance of economy and market leads to an unworld, a world which is not inhabitable. Instead we now need a new form of globalization which would lead to creation of the world. . . . For Nancy (2007), while globalization is "uniformity produced by a global economic and technological logic . . . leading toward the opposite of an inhabitable world, to the *un-world*," *mondialization* involves authentic world-forming what Nancy calls creation of the world. Cultural creativity and regeneration is at the heart of such alternative creations of the world.

9 It is helpful here to know the work and thoughts of Nicanor Perlas a bit. Perlas has met with Rajagopal and both of them have met in Europe under the auspices of groups and individuals drawing inspiration from Rudolf Steiner such as Dr. Karl-Julius Reubke. Nicanor Perlas himself draws inspiration from the work of Rudolf Steiner.

10 This raises the question of what kind of education and pedagogy of education about which Gandhi, Sri Aurobindo, and Steiner have a lot of important insights to share. In WSF meet in Mumbai in 2004, Dalit groups had made their presence felt. Now Dalit groups and social movements are becoming more active in India. They refer to Ambedkar's Trinitarian call for social change: organize, educate, and struggle. But what kind of education Dalits and all of us need to have remains not much explored. Here both Steiner and Sri Aurobindo cultivate pathways of multi-dimensional learning such as educating our body, mind, psyche, vital, and spiritual. Exploring visions, practices, and pathways of education is an important challenge before us. This, of course, calls for far deeper cross-cultural dialogues across paths of strivings and experimentation rather than just reproducing thoughts and perspectives of Ambedkar, as Dalit movements and intellectuals have been mostly doing.

11 Thinkers such as Jurgen Habermas invite us to go beyond conventional secularism and nurture creative post-secular moves. But though this points to a global spiritual heritage, Habermas and fellow friends from the Euro-American world do not do this enough. In places like World Social Forum and Parliament of World Religions, we get pathways of open post-secular experiments which build upon multiple philosophical, religious, and spiritual traditions of humanity.

12 This is retrieved from the web.

13 Manuel Castells from www.irishleftreview.org/2012/01/27/indignados/#sthash.q7huNNJY.dpuf

14 This section of Indignados builds upon the research and text written by my research associate Maanya Rao for this book project.

15 By simultaneously engaging with Gandhi and Ambedkar in the Indian context, Rajagopal and Ekta Parishad are performing an important hermeneutic task of social liberation. The same is true of their efforts to bring Dalits and tribals together. This bringing together happens not in moments of stasis but in dynamics of movements such as Dalits and Adivasis marching together. It is an important aspect of a multi-topial hermeneutics in contemporary India, the vision of which was outlined in the first chapter of this study.

16 Rajagopal has taken part in an international conference on Alliance Building for a New World nurtured by Goetheanum in which he had titled his presentation as "Wanting to Suffer More Than Others." In this, he had also said:

> Our responsibility to act swiftly and collectively is great. No individual can work on their own to have any impact against the effects of globalization. Forming alliances and keeping them alive demands sacrifice and willingness

to learn from each person. Different groups, different organizations have different abilities: how do we bring them together?

17 Muhammad Yunus, Nobel Laureate, and much known founder of Grameen Bank in Bangladesh also challenge us to realize this in his work and in his thoughts (see Yunus 2010).

18 Rajagopal shared this with the author in a recent discussion with him at Jai Jagat Center, Puducherry, in April 2021.

19 Radha Behen, Radha Bhatt, was the President of Gandhi Peace Foundation. She herself is an inspiring activist and thinker having nurtured Lakshmi Ashram in Kaushani and having studied and worked with the legendary follower and co-worker of Gandhi Sarala Behen. During Jan Satyagraha 2012, Radha Behen was walking with us for the first few days. One day, she told fellow walkers and Rajaji:

> Rajaji and some of you have walked across the whole country. You have done your *tapa*. The fire of this *tapa* gives you strength.

20 Jan Satyagraha 2012 passed through a small town near the Ashram. I visited one brother associated with this Ashram and stayed with him. I visited the Ashram. This was a place where Rajaji stayed for sometime during his early days of Gandhian creative social action. At this period, he worked closely with Dr. Subba Rao who continues to be a guiding light to him as well as Ekta Parishad.

21 Forest Rights Act of 2006.

22 About this phase of his life and work in Chambal valley, Rajagopal tells us in an interview:

> I wanted to use non-violence very effectively. Some people are impatient. They take to the gun. I am saying before you opt for the gun there is one more possibility.
>
> (Drakakis and Williams 2005: 109)

23 Ramesh has then written a book on the new land acquisition act which he helped to draft (Ramesh and Khan 2013).

This clause allowed the collector to "take possession of the land within 15 days of giving notice." He could take possession of a building within 48 hours of giving notice. "The Outer Ring Road Project of Hyderabad and the Expressway in Uttar Pradesh are both striking (and recent) examples of acquisitions where large tracts fell prey to the urgency clause," write Ramesh and Khan.

Furthermore, land acquisition displaced many people over the decades and most of them were not resettled and left to fend for themselves. Ramesh and Khan write:

> While there is no comprehensive record of how many individuals have actually been displaced by land acquisition post-independence, estimates put forth by credible studies find that close to 60 million individuals have been displaced since independence. Worse still, only about a third of these have actually seen some measure of resettlement and rehabilitation.

Furthermore, the studies that Ramesh and Khan refer to are more than a decade old. Hence, the number of displaced is likely to be higher than 60 million.

24 According to Dr. Reubke, Ekta Parishad played important role in this resistance but it did it as part of an alliance and did not want to take single-handed credit for this.

25 In her presentation in a discussion at Jawaharlal Nehru University (JNU) on April 18, 2016, Bela Bhatia, a scholar and activist, presented the situation in Bastar in this way.

26 This emerges out of a discussion I had with Kankar in JNU in April 2016 when he was visiting JNU.

27 During our discussion in JNU after his presentation on April 18, 2016, Komal Kanker told us about this.

28 Discussion with Professor Mohanty at his office at Council of Social Development.

29 Nachika Linga fought along with people of his village and occupied land in Narayanpatna. He was branded a Maoist. He had absconded to avoid police arrest, but after a while, he is now in jail.

30 In a recent note on the vision and the practice of *Shram Bank*—Labor Bank, Rajagopal (personal communication) writes:

> As a daily wage earner the only asset I have in my control is my physical labour. I am willing to sell my body labour every day in-order to buy my ration to keep my family going. The problem begins when people who are willing to sell their physical labour find no market to sell it. If I don't sell my labour today, I can't keep it for tomorrow. My body energy of today can't be reserved for tomorrow.
>
> If you have money that you don't want to use today can be deposited in a bank and you can draw when you need it. Can we see labour as money or wealth and allow him or her to deposit it in a labour bank? Can one deposit the body labour against grain so that the family will not go hungry?
>
> This is a fact that the grain bank can't provide work but it can guarantee food grain against the calculated labour of 8 hours and can provide food grain for an equal amount of money. (Say Rs 200/- is the cost of my labour for a day, so I can deposit it against 10 kgs of grain) and against an undertaking that when I am able to sell my labour I will return this grain to the Bank. This is also possible that a particular person has no work for a week and he/she would like to make an undertaking for a bigger quantity of grain with a commitment to return when work is available.
>
> We are also exploring the possibility that the bank can generate work projects that can create community assets or the bank can at least keep a record of unemployed people and negotiate with projects where manual labour is in demand.

31 I have discussed practical mysticism and spirituality and its relationship to politics elsewhere (see Giri 2013, 2020b).

32 As Doran also writes:

> In recent years, a new trend has been emerging: the study of the incorporation of new religious dimensions such as blessings, collective prayers, or other religious rituals in the social practices of popular and social movements in Latin American countries. In Brazil, scholars are turning attention to the importance of *mistica*, namely self-constructed and self-conducted religious celebrations held at every meeting of MST.
> (Doran 2014: 234)

3 New Initiatives in Dialogues Across Borders

Transforming Theories and Practices of Dialogues and Responsibility in India, Indonesia, and the World

Dialogue is a process of prayerful mutual encounter, and to reach out to the other in a spirit of dialogue calls for a commitment to truth, and also a growing awareness of the fact that truth cannot be "possessed" in the strict sense of the term, and that through dialogue we become co-seekers of truth. Besides, there is no question of knowing the Absolute Truth, for Man as long as he/she is a traveller on this earth is bound by the conditions of this world and cannot transcend them howosever it may be desired. Hence, all claims to Absolute Truth are a myth and what we can try and search for is the truth it its in-finiteness. In this constant search what is granted to us is to be partners without domination, and also respect the other as a source of truth. Truth although . . . is multi-faceted and we can only have a glimpse of some aspects of the truth, as the world famous story of the blind men and the elephant indicated. We are conditioned by our perspectives, attitudes and culture; and what is needed is that we accept these limitations and cooperate with others in the search for Reality.

—Raimundo Panikkar (2008), "Introduction"
Toward Mutual Fecundation and Fulfillment of Religions, p. 15

A prayer which asks for nothing, and receives
nothing? A prayer which
dare not acquire
a sound,
dare not resound
in the house,
dare not become a weeping?

Rustam Singh (2011), "Weeping" and
other Essays on Being and Writing, p. 101

Dialogue is an inescapable part of our life, but it is, at the same time, a continued challenge. Dialogue deals not only with the self and the other but also with the world. Therefore, dialogue is not just dialogue; it is part of a broader multi-logue and polylogue. Dialogue involves words and worlds in complex ways. Dialogue involves communication but it also embodies love, labour, mediation, and meditation. Dialogue is a multi-dimensional and transversal

DOI: 10.4324/9780429347481-3

journey which involves conflicts and strivings to overcome conflicts, compassion, and confrontation.

Dialogue and responsibility are integrally linked. As there are growing movements of violence and hatred and monological closure of many kinds, there are also many movements of dialogues across borders. In this chapter, we discuss some of these initiatives.

We begin this with a description of some initiatives in inter-religious dialogues in two countries of our present-day world: India and Indonesia. Both of these are multi-religious societies. While majority of Indians are Hindu, majority of Indonesians are Muslims. But despite demands from certain sections of majoritarian groups for a religious state, both the countries are secular. India is a secular state nurtured not only by Constitution but also by traditions of spiritual opening to other religions and mutual learning which has existed for millennia. Indonesia is a secular state which is guided by the State ideology of Panchasila, but despite this being a state project, it also has its deep roots in traditions of co-existence and openness to each other in Indonesia. For example, Islam in Java has embodied much more border-crossing with other religious streams especially Hindu streams from India (cf. Geertz 1960).

With this, let us look at some of the initiatives in dialogues in these two countries. We begin with Indonesia

Indonesia

Indonesia is a multi-religious society. Though Panchasila has provided a secular foundation to state, it has not permeated the whole of society and there has been historically and contemporaneously political efforts to Islamize it and give more privilege to Muslims. As we would see, thinkers, leaders, and statesmen like Abduraddin Wahid have opposed such Islamization moves but nonetheless it has continued. There has also been violence between Christians and Muslims and places of worship from both religious communities have been attacked and burnt though in many cases Christians being in minority have borne the brunt of the attack.

At the same time, there have been many initiatives in dialogues. Wahid was the leader of Nahdlatul Ullama (NU) and it always promoted a tolerant view of Islam. In Salatiga in Central Java, Percik, is a research institute and a voluntary organization working on dialogue and democracy. It has created initiatives in grassroots dialogues between Christians and Muslims with participation of Buddhists when needs be. I had done field work with Percik for a week in Salatiga and surrounding areas in February 2015. But before I describe the work of Percik in the field of inter-religious dialogue, it would be helpful to have a view of cross-currents of inter-religious dialogues especially between Christians and Muslims in Indonesia. For this we can build on a doctoral work on this subject by Zainul Fuad. In his doctoral thesis, *Religious Pluralism in Indonesia: Muslim-Christian Discourse*, submitted at

University of Hamburg in 2007, Fuad (2007) discusses the background of state and society in Indonesia especially the ideology of Panchasila and some initiatives in dialogues initiated by both Christians and Muslims.

Through Panchasila, Indonesian state strives to create inter-religious harmony. The council of Indonesian Ulama plays significant role in managing inter-religious relations in Indonesia. At the same time, there have emerged other initiatives in inter-religious dialogues such as INTERFIDEI (Institute for Inter-Faith Dialogue in Indonesia). This was created in 1992 by leading Indonesian Protestant thinker The Sumathana. Since 1998, their local conflict resolution workshops have "both an inter-faith and inter-ethnic perspective."

Fuad discusses the thoughts and works of some Islamic thinkers and leaders in Indonesia who have worked on dialogue between Muslims and Christians. He begins with the work of Nurcholis Madjid. Madjid urges all concerned to realize that Prophet Abraham or Ibrahim surrendered himself to God but was not committed to a certain form of "organized religion" (Fuad 2007: 129). Madjid tells Muslims to realize that Prophet Muhammad had developed the Medina Charter so that Muslims and non-Muslims are united within a band of civility.

Abdurrahman Wahid was a pioneer of inter-religious dialogue in Indonesia. Wahid was known as Gus Dur. Wahid developed *pesantren*, schools of education working in Indonesia which are like Madrasas, as places of inter-religious dialogues. In 1991, Wahid, together with leading figures from different religious backgrounds, set up a forum called *Forum Democrasi*. Wahid was a courageous Islamic thinker and he argued that Indonesian Muslims should just say Salamat Pagi or Salamat Malam and not Salam Walakum which would be a reflection of localization of Islam. Wahid opposed Islamization of state by the formation of ICMI. Wahid criticized the issuance of the 1989 Law of Religious Education and this Law, according to him, would be used by *dawa* groups to promote narrow "Muslim-only" concepts in the school system. For Wahid, Islam should be "positioned as a complimentary factor, and it may not dominate the life of society" (ibid: 143). Fuad tells us, "Like Madjid, Wahid also emphasizes the significance of Panchasila as a principle that should be maintained in inter-religious relation" (ibid: 145). "To Wahid, the acceptance of Panchasila is the consequence of the relation between Islam and the state based on Islamic consideration" (ibid: 146–147). Wahid was the President of Republic of Indonesia from 1999 to 2001, and as President he continued his efforts at inter-religious dialogues and co-existence. A dialogue between Abdurrahman Wahid and Daisaku Ikeda has come out as *The Wisdom of Tolerance: A Philosophy of Generosity and Peace* (Wahid and Ikeda 2015). Wahid also founded Wahid Institute, and in this book, he tells us Ikeda that through the activities of the Wahid Institute, "we are working to promote the harmonious co-existence of different religions and win broader acceptance of cultural diversity" (Wahid and Ikeda 2015: 3).

Wahid says that originally Sharia meant that it is a Way rather than a law. He also urges us to realize that laws of blasphemy and apostasy were formulated in a political context and it should be abandoned. Speaking of the Indonesian context, he says, "if this becomes so then we would have to execute 25 millions of Indonesians who have changed religion."

Banawiratma is a Christian theologian who emphasizes the "significance of the teachings of the Second Vatican Council as the basic foundation for the opening of Catholic attitude in relation with other religions" (ibid: 172). He finds it necessary to "develop another level of dialogue, which he calls 'contextual analysis and reflection'" (ibid: 176). He elaborates such a practice of dialogue:

> This level of inter-religious dialogue takes place in small groups who know each other, in daily life where men and women of different faiths experience together a common situation, with ups and downs, anxieties and hopes, and thus common concerns emerge. They are concerned about the need for clean water, healthy housing, adequate education, fields of work etc.
>
> (ibid)

Expressing solidarity with Muslims, Baniwiratma says, "We, Christians, address the same God as Abba, the motherly Father of Jesus and our motherly Father." It may be noted that Baniwiratma presents God as Motherly Father which already embodies a deep realization of not only God as Mother but also Father as Mother. Baniwiratma says that the meeting point between Christians and Muslims is Word of God and not the book of God such as scriptures. While for Christians, Jesus is the Word of God, for Muslims, it is Quran. So for Baniwiratma, the comparison between Christians and Muslims is not between Jesus and Muhammad but between Jesus and Quran.

Franz Magnis-Suseno is originally from Germany and has got settled in Indonesia. He is a respected Protestant theologian and public intellectual of Indonesia. He emphasizes on natural law as the foundation for dialogue. He also emphasizes on ethics as a foundation of inter-religious interaction and dialogue. "Like other Christian theologians, Magnis-Suseno also emphasizes the significance of Panchasila in dealing with the plurality of Indonesian society" (ibid: 194).

Eka Dharmaputera is one of the founders of Institute for Inter-Faith Dialogue in Indonesia. Dharmaputera has an open way to Christian path and the way of Jesus. As Fuad tells us, "Dharmaputera indeed acknowledges the absoluteness of the claims of Jesus as the Way to the Father. However, he remarks that this way is never identified with a particular religion" (ibid: 203). "Jesus is the way, not religion,' he says" (ibid: 203). "Evangelization is to make Christ known not to make our religion bigger" (ibid: 205). He considers Pachasila as crucial to inter-religious harmony, but he says, "Panchasila should be realized not only in state life, but also in the social life" (ibid: 206). Like some of other inter-religious dialogue partners in this journey, he brings a deeper engagement with Panchasila: "Panchasila should even be done in the form of obedience to God" (ibid: 205).

Percik and the Work of Inter-Religious Dialogues in Indonesia

Percik is a research Institute and voluntary organization working on local democracy and dialogue in Indonesia. It was founded by Dr. Prajdarto, who did his doctoral work in Anthropology from Free University, Amsterdam. Pradjarto had studied the work of Pesantren and Nadthtula Ulama (NU) for his doctoral work. His study of NU and close association with Gus Dur had made him interested in both democracy and dialogue. He founded Percik as an Institute to work on both these interconnected streams. It began during the last era of Soheartao regime when there was still a lot of dictatorial control on any free initiative. Pradjarto and his associates were closely watched upon sometimes spies parking in front of his home. Percik had a humble beginning with the love and labour of Pradjarto and his co-enthusiasts. During our conversation, Pradjarto was telling us that, even with a doctorate in anthropology, he was trying to sell rice in the market place to generate resources from this sprouting seedling.

To work as an independent research Institute and voluntary organization is not easy in any country more so in Indonesia where there seems not to be a culture of support of Government for such initiatives and there is not also a widespread culture of private philanthropic support for such work. After 25 years, Percik continues to have financial challenges, but Pradjarta is hopeful that it would continue its work and is creating capable young leaders to continue.

I had first become interested in the work of Percik when I had visited it in December 2003 while co-organizing an international conference on Development with my dear and respected friend Philip Quarles Van Ufford from Vrije University (VU), Amsterdam. Pradjarto had studied with Philip in VU, and when we published an edited volume on *A Moral Critique of Development*, Pradjarto became a mid-wife to its translation in Bhasa Indonesia. We held an international seminar on Ethics of Development as a part of release of this Indonesian translation of our book in December 2003. After the seminar, I stayed for a few more days in Percik and visited the villages where Percik works. While visiting a village where Percik works on local democracy, Percik staff told me how they also organize dialogue between Muslims and Christians in villages. I had all along been interested in inter-religious dialogue but had met few instances of inter-religious dialogues involving common people in grassroots level and obviously I became much interested to learn about it further. Being in this journey also inspired me to link dialogue with responsibility which then blossomed into the current research project.

After our initial meeting in 2003, I visited Percik in February 2015 and spent a week there. During our first meeting, Pradjarto told me about

Percik's many-sided vision and activities in inter-religious dialogue. It organizes Sobath (friendship meet) between Muslims and Christians. It also organizes meeting between Islamic teachers (*Kiais*) and Christian priests (*panditas*). Once Percik had organized a meeting where 15 *Kiais* and 15 *Panditas* stayed together. It was an occasion to discuss fears and apprehensions from each other's sides. Muslims have a long-standing fear and apprehension that Christians are engaged in conversion with the superior power of money and with the benefit of Dutch colonialism. Christian lay people and *panditas* have also fear and discrimination that Muslims have the power of muscle and majoritarian state forces behind them. They also believe that Muslim young men deliberately entice young women from Christian community and sometimes impregnate them with the sole purpose of converting them to Islam, an allegation which also resonates with similar allegations from Hindus towards Muslims in both India and Britain. But the Sobath meet was an occasion to openly discuss each other's fears and apprehensions. After this meet, one Pandita remarked, "If I tell our people that we have become friends, they won't believe in it." But the Sobath meet was not just a one-time event. Percik organizes regular Sobath meets not only among the priests but also among ordinary people. It has also Sobath meets separately for youngsters, which is called Sobath Muda and for children which is called Sobath Ana.[1]

Percik also uses the traditional Javanese practice of *Sambartan* where people in a village come together to work together with their homes and fields such as working together in agricultural field or repairing and build each other's homes. Percik builds upon this traditional Javanese practice and brings people of different religions together to build houses of prayers. Such local resources for inter-faith work are quite crucial. At a more fundamental level, during our conversation, Pradjarto urges us to realize that the "problem of inter-religious dialogue is one of universal theology. Muslims and Christians wanted to keep the Western and Arabic version of their doctrines. What is needed is contextual theology." As we saw in previous paragraphs, both Christians and Muslims have been engaged in creative contextual theologies and local spiritual work with their religious traditions. Such localization is a help in inter-religious dialogue.

During my visit, there was a programme on Inter-Religious Dialogue in the local Pesantren. The students and teachers of the local Pesantren had taken part in this meeting. The Pesantren is run by a Kai who belongs to NU and is therefore more dialogical and open to other religions carrying the spirit of Wahid. But the speakers of this seminar had a more restricted view of inter-religious dialogue. For them, inter-religious dialogue and religious pluralism do not imply that all religions are same. A speaker said, "We accept multi-culturalism at the social level and not at the religious level." He went on, "Pluralism is a concept to destroy Islam as the West failed to do it through secularism. Exclusivism is needed to protect Islam from other religions. Islam is 100% true but it can still be respectful of other religions."

Figure 3.1 The author with Agung, Heru, and Ambar—friends working with Percik.
Source: Author's Archive

After attending this lecture, I had a discussion with the founder of the Pesantren. As already mentioned, he has a much more liberal approach to inter-religious relations. About contemporary Indonesia and the challenges of inter-religious conflicts and harmony, he said that there are conflicts, but they are because of economic resources. About the relationship between NU and Muhammadiya (another major Islamic group in Indonesia), he said that they are now trying to solve common problems together. For him, "The fight between NU and Muhammadiya can be solved at the social level. The approach of the Pesantren is much wider. We need to solve the problem of relationship between religions in practice." He has been associated with the dialogue work of Percik and he appreciates its work. Both Percik and the Pesantren help each other in creating a ground for inter-religious learning and dialogue. Percik often brings outside visitors to this Pesantren, as it also brings participants in inter-religious dialogues to experience life in Pesantren.

There is also a programme of Junwadi which focuses on the role of Islamic schools in preventing conflicts. It involves exchange of students from one *Pesantren* to another.

Pekalogan is a seaside city in Java where people of many different ethnic and religious backgrounds stay. I visited this city to understand dynamics of co-existence and works in inter-religious dialogues. I was accompanied by Agung, son of Pradjarto and a worker with Percik, to Pekalogan. Agung is also completing his master's on social capital and inter-religious social

Figure 3.2 The author meeting a local leader.
Source: Author's Archive

networks in Pekalogan. We visited a local priest of a Christian church, a charismatic Islamic leader, and a leader of Indigenous spirituality. All the three of them are committed to inter-religious dialogue.

Let us begin our journey of learning with Maulana Muhammad Habib Luthfi, an important religious leader in the city. Lutfi was born on November 10, 1947. He is an important leader of Islamic community in Pekalogan and also Java. He has grown up in Nahdatula Ulama and thus shares its open-ended approach to inter-religious relations. About his experience, he says, "I was a leader of youth wing and was associated with the paramilitary wing providing security to both NU and other religious groups such as Christians. When churches are attacked, we provide them security."

Luthfi is open to other religions, but when I asked him if he studies Quran and Bible together, he said no. He runs an NGO to help people and solve social problems. He is particularly focusing on "religious rules which can protect women." On behalf of his NGO, he also organizes Sobath meets between people of different religions, especially Christians and Muslims. But during Sobath meetings, "We do not have a set agenda. We talk about social problems, human rights issues."

After meeting with Luthfi, we came to meet with Asworo Palguno who is trying to revive indigenous Javanese spirituality. He, along with his wife, runs a restaurant which is also a meeting place for indigenous spirituality. It has symbols of indigenous spirituality and many yellow flags which is the colour of indigenous spirituality.

Aswono is trying to revive indigenous spirituality of Earth. During our meeting, he mentioned that he is planning to organize a meeting near Solo to pray for Earth spirituality. He said:

> We can communicate with soil. I took some soil from Percik. Soil is a medium of communication. . . . To unite the Earth from everywhere. If we unite the earth, we can make peaceful living. People cannot live without Earth. Earth is our Mother. Our Mother Earth is crying because there is a fight between her sons.

Reviving indigenous spirituality in Java is difficult and needs practice. He teaches his children and other children with the practice of his own spiritual living. In his words, "Religion without comment. People do what is in their heart. In a horizontal relationship children learn by watching what the other does. The other has to be wise in doing."

The challenge of inter-religious learning and dialogue is on both sides and it is a responsibility for all concerned. He said that there are also many in indigenous Javanese religions who do not like Islam: "How it is different? I live in a plural society. My father told me to learn about and respect other religions. I learnt about Islam from my wife that is not a religion but a spiritual practice." The crucial issue vis-à-vis Islam and other religions is the place of local culture and roots. In his words, "We cannot take our local reality from us. But they are people who would like to go out of Java and become Arabic." What he is suggesting is that for inter-religious harmony and dialogue, creative localization of one's religious belief is crucial, a thought which also resonates with many thinkers and practitioners including those of Pradjarto.

During this visit, I also met a leader of a Pesantren who is a young man. He is trying to make this Pesantren a Green Pesantren by encouraging students and staff to green the local area and clean the local stream which is polluted by chemicals from the Batik industry. He is teaching students of his Pesantren who are known as Santris simultaneously grammar and ecology. When I asked him if there is a connection between grammar and ecology, he answered in the affirmative: "Learning grammar makes you more ecological." Traditionally Pesantren has not paid enough attention to the teaching of science and technology and he is planning to do this in his school. On inter-religious teaching and learning in Pesantren, he does not feel any special need for it. He feels that to teach about religion in Pesantren is to ask them to feel the difference of the other, other religions. He is suggesting that students can do this on their own if they want.

We had a meeting with a leader of Christian community before our departure. About inter-religious dialogue, he said:

> Dialogue can be done if there is friendship. We need friends to dialogue with as religion is a very sensitive issue especially in Indonesia. Religion

is both *bichara* (belief) and *achara* (conduct). If we talk about *bichara* (belief) there is conflict but if we focus on *achara (conduct)*.

We also talked about Panchasila. I asked him if Panchasila promotes dialogue among religions. He said, "Panchasila is to protect each religion." But he did not elaborate whether Panchasila creates dialogue among religions. He said, "Religion protects Panchasila. Faith is not talked about. Can a Buddhist receive the concept of *Trimurti?*"

After this, I visited the Islamic University in Semarang (UIN Walisongo Semarang) where I presented a lecture on "Inter-Religious Dialogue and Global Responsibility." Professor Mushadi of the university was my host. He and his colleagues and students were open to discussing fundamental theological issues. For instance, in my presentation, I spoke about the need to build on transformation of both science and religion for the purpose of inter-religious dialogue. I also said that in Islam, we always begin our prayer to *Bismilla Al Rahiman ir Rahim* which opens us to God the Merciful. But in organizations of religion including in organizations of inter-religious dialogues, we do not cultivate enough God the merciful. We reproduce the logic of God the powerful. But the fundamental challenge for inter-religious dialogue is to cultivate our own realization of God as Merciful. We also need to cultivate our own realization of God as Mother. But though Allah is gender neutral, in conventional Islam as in the broader Abrahamic tradition, there is a patriarchal binding of God as Father which today needs to be also realized as Mother. In realizing the Motherly aspect of Allah in Islam and God in Christianity, inter-religious and trans-religious spiritual journey with other religious and spiritual traditions where God is realized as Mother such as the Mother Goddess traditions like Tantra would be helpful.

What was amazing that there was a deep opening to these streams of thinking on the part of Professor Mushahadi and his colleagues and students. The next day I also presented a lecture on inter-religious dialogue and literature at Islamic University Salatiga in a class. In my lecture and interaction with the students and class professor, I said that now we need to bring the engagement of creative literature to our practices of religions and inter-religious dialogues. Religion has a poetic dimension which is much more fluid compared to the dimension of prophecy and belief. If we relate to our own religions and discover its religious and spiritual meaning through poetry, we touch the universal vibration of Truth in our own religion which then also opens us to this universal vibration and rhythm in other religions. When we learn about other religions if we begin with their poetry, it can also touch us with and our belief and claims to Truth and certainty. After our discussion, I invited students to dance together in the class by forming circles of pairs. I said that we can also chant Salam Walakum and Aum which is a universal sound together. In Jakarta and many other places, I have also danced this chanting of Aum and Salam Walakum together.[2]

If, in my discussions with students and colleagues at Islamic universities, I had discussed the way we can cultivate inter-religious dialogues with the poetic creativity of our lives, I had not discussed about the way it can also be cultivated through creative films. In the middle of night one day, an Islamic scholar from Jakarta brought me to this realization. One night, we were having a dinner with Mr. Suswono, the former Agriculture Minister of Indonesia. Our friend who is a teacher in a Pesantren and a Kiai was with us. After our dinner, he put me on the back of his motor bike and brought me to my hotel. On the way, we stopped in a place. He told me about his love for Indian movies, especially Bollywood movies, which is common among Indonesians. He told me that, for him, movies are a great way of teaching about religion and about inter-religious dialogues.

On being back to our hotel, we sat together and talked about the challenge of inter-religious relations in Indonesia. He told me that in Indonesia, there is a struggle between the hardliner and tolerant Islam. He loves the work and legacy of Gus Dur. But many of his friends even do not want to hear about it. But he always pleads with them and challenges them to be open to the work of savants of dialogue and harmony such as Gus Dur.

At the same time, Indonesia is a place of many creative transformations of the religious domain. One of this relates to spontaneous reworking of gender relations. Usually in Islamic places of worship, women pray in a separate room. But I saw in many places women and men praying without much of veils and walls of separation. At UNPAR, Bandung, a Catholic University, there is a makeshift place of prayer by a staircase. On one side of this place of worship, men pray, and at another side, women in the same place and upon the same mat pray on the other side. Similar is also the situation in a makeshift place of worship near the main historical museum in the heart of Jakarta. The big mosque in Jakarta also does not have wall of separation between men and women in the big prayer hall.

Dialogue of Civilizations

Clash of Civilizations has been the regnant discourse in the last decades with the publication of an article and a book with the same title by Samuel Huntington (Huntington 1996). Thinking with this and moving beyond, many seekers and movements around the world have taken the initiatives towards dialogue of civilizations in discourse and practice. Fred Dallmayr, an important philosopher and thinker of our present-day world, wrote a book *Dialogue Among Civilizations* in which Dallmayr (2002) tells us how civilizations have been in dialogues with each other through out histories and societies. Years later, Richard Harz, a seeker in the path of integral spirituality of Sri Aurobindo and The Mother, wrote a book called *The Clasp of Civilizations: Globalization and Religion in a Multi-Cultural World* again showing us ways of going beyond the discourse and practice of clash of civilizations (cf. Hartz 2014). For Dallmayr, dialogue among civilizations involves not only human beings but also Nature and Divine. For Dallmayr, cross-civilizational dialogues cannot

> remain entirely human-centered or polis-centered. Despite the importance of civility and civilizational discourse in "cities" (the etymological root of "civilization"), a genuine cross-cultural meeting has to take into account the deeper dimensions and resonances of human experience, differently phrased, it has to make room for certain corollaries or supplements of civilized life—corollaries that are thematized here under the rubrics of "nature" and "divine."
>
> (Dallmayr 2002: 3)

Appreciating this turn to Nature in Dallmayr's dialogue among civilization, John Clammer, nonetheless, urges us to realize that we need to relate this to "organic interconnection to the common ground of our being-in-nature" (Clammer 2016: 126). Cross-civilizational dialogues ought to take place on the ground of Nature. For example, cross-cultural dialogue can focus on climate justice. This calls for cross-cultural dialogues on developing ecological self in us. Cultivation of ecological citizenship is much more than global governance or planetary citizenship as it involves "self-transformation of citizens without which institutional changes are at best superficial" (ibid: 130).

Dialogue among civilizations is part of a movement towards a new civilizational transition and building a new civilization going beyond anthropocentric dominance and a new civilization in which Nature, human, and Divine play creative roles. Here we can bring the works of Fred Dallmyar, John Clammer, Arturo Escobar, and Candido Grzybowski together and link dialogue among civilizations to civilizational transitions and building a new civilization, a bio civilization. Cross-cultural dialogues should help us building a bio-civilization where we "become reintegrated with life and the dynamics and rhythm of ecological systems, adjusting to them, enriching them and facilitating their renovation and regeneration" (Grzybowski 2019: 102). We can link dialogues among civilizations to a new civilizational transition. As Escobar writes:

> The notion of civilizational transition establishes a horizon for the creation of broad political visions beyond the imaginaries of development and progress and the universals of western modernity such as capitalism, science, and the individual. It does not call for a return to "authentic traditions." . . . [I]t adumbrates pluralistic co-existence of "civilizational projects" through inter-civilizational dialogues that encourage contributions from beyond the current Eurocentric world order. It envisions the reconstitution of global governance along plural civilizational foundations, not only to avoid their clash, but to constructively foster the flourishing of the pluriverse.
>
> (Escobar 2019: 123)

In the field of international politics and world engagement, President Muhammad Khatami of Iran made dialogue among civilizations a key theme in world discourse and he and Iran played an important role in UN declaring 2001 as a year of dialogue among civilizations. Khatami had studied political philosophy and had worked on Aristotle. He taught at Tarboat Modannes University in Iran. Following the earlier work of noted Iranian philosopher Dariush Shayagen, Khatami introduced the theme of "Dialogue Among Civilizations" into the UN discourse. He held a famous interview with Christina Armapour in which he emphasized the reality and need of dialogue among civilizations such as American civilization and Iranian civilization. With deep philosophical and historical sensitivity, Khatami tells us in this interview how the foundation of American civilization lies in the Plymouth Rock planted by the pilgrims which is a symbol of combining religion with liberty. This is also for him the core of aspiration of Iranian civilization which is a confluence of Persian and Islamic civilization.

It may be noted here that Khatami grew up in the city of Yazd which was one of the ancient centres of Zorastrianism. Zorastrianism has been one of the earliest open religious and spiritual explorations of the world which has inspired critical and dialogical reflections down the centuries in such philosophers as Frederic Nietzsche and later day religious and spiritual formations such as the Bahais which have been much more open to all the religions of the world. Khatami may have drawn upon his native cosmopolitan culture of openness and deep learning. In his address to the UN Year of Dialogue Among Civilizations in 2001, Khatami tells us:

> What we ought to consider in earnest today, is the emergence of a world culture. World culture cannot and ought not to overlook characteristics and requirement of native local cultures, with the aim of imposing itself upon them. Cultures and civilizations that have naturally evolved among various nations in the course of history are constituted from elements that have gradually adapted to collective souls and the historical and traditional characteristics. As such, these elements cohere with each other and consolidates within an appropriate network of relationships.
>
> (Khatami et al. 2001)[3]

Khatami founded Foundation for Dialogue Among Civilizations in 2007 whose European headquarters is in Geneva. It collaborates with similar initiatives such as The Oslo Center for Peace and Human Rights, The Washington National Cathedral, Foundation Culture of Peace, and Club de Madrid. All these partnering organizations are committed to peace and dialogue. For example, Foundation Culture of Peace was founded by Frederico Mayor, the former Secretary General of UNESCO. Mayor began as a scientist but became a politician and statesman and he is also a poet. Probably the poetic spirit of Mayor dances with the philosophical spirit of Khatami and together along with partner organizations they strive to create more streams and spaces of dialogues in this fragile world. It may be noted that when the Alliance of Civilizations began as an initiative of UN in 2005, Kofi Annan appointed President Khatami as a member of this. In 2009, he shared the Global Dialogue Prize with his philosopher co-pilgrim Dariush Shayegan from Iran. While his Foundation for Dialogue Among Civilizations works at an international level, Baran Foundation established by Khatami works in Iran on domestic issues of well-being and dialogue.

Anthony J. Dennis wrote a booklet in 2004 called *Letters to Khatami* in which he appreciates the courageous step taken by Khatami in moving beyond the rhetoric of clash of civilizations.

United Nations Alliance of Civilizations[4]

The idea for an *Alliance of Civilizations* was first put forth in a speech to the 59th session of the United Nations General Assembly given on September 21, 2004 in New York by the former Prime Minister of Spain, José Luis Rodrí-guez Zapatero. Just before half a year, on March 11, a terrorist attack had taken the lives of 192 people in Madrid. Zapatero reminded those gathered how the men and women from Spain "took to the streets and squares of the cities," expressing "our rejection and disgust, our unanimous contempt for terrorist brutality." He also expressed sympathy for, and solidarity with, not only those in the USA who suffered great loss during the 9/11 terrorist attack, but also all those in other locations globally, including Jakarta, Bali, Casa-blanca, Riyadh, and Beslan, who had been afflicted by terrorism throughout recent years:

> Here in New York I would like to convey the Spanish people's heart felt solidarity with this great American nation. We fully understand the ter-rible pain that in these past years has been inflicted. We know all about enduring kidnappings, bombings and cold-blooded killings. We are well acquainted with the meaning of the word compassion.

Zapatero raised serious concern, however, about the nature of the US response to terrorism and what its implications were for democracy, interna-tional law, and the rise of extremism. Speaking from his direct experience of terrorism in Spain, he reported, "we have learned that the risk of a terrorist victory rises sharply when, in order to fight terror, democracy betrays its fun-damental nature, governments curtail civil liberties, put judicial guaranties at risk, or carry out pre-emptive military operations." Zapatero's great hope in the struggle to overcome the irrationality of terrorism, which he likened to "the Black Death" and the plague, was the rationality of law itself. He affirmed that "legality, democracy and political means" make nations stron-ger, and that "legality, and only legality" would allow the world's people to prevail in the fight against terrorism.

He added that Spaniards would resist terrorism, but always "within the framework of both domestic and international legality" by respecting Human Rights and commitment to the United Nations, "and in no other way." In support of a new kind of solidarity, one built on a sober optimism, that expressed greater faith in the capacity of humankind to practice high reason and principled reflection, he declared, "It is not only the ethics of our beliefs that move us, but also and above all our belief in ethics."

He further expressed the hope that peace and security would spread over the world when the foundations of just democracy are established. In an affirmation of soul force, truth force, moral force, and compassion, he concluded that ultimately the strength of dialogue itself, among and across peoples and civilizations, would ensure our true progress as a human family.

Zapatero's powerful communication of his "belief in ethics" and in the strength of dialogue related directly to discourse emerging from the World Public Forum (WPF), an initiative involving civil society representatives from Russia, India, and Greece which organized the International Programme "Dialogue of Civilizations." WPF "Dialogue of Civilizations" was accepted on November 9, 2001, by former President of the Islamic Republic of Iran Mohammad Khatami, and was a practical realization of the UNGA resolution "Global Agenda for Dialogue among Civilizations," discussed in the previous section. In a speech delivered by Professor Fred Dallmayr at the 9th Rhodes Forum Session, in October 2011, Dallmayr described the values that have defined WPF from the outset, saying, "what we most intently are for is a world community in which societies and civilizations interact in the spirit of mutual respect and dialogical cooperation." He connects the initiation and purpose of WPF with the most profound aspirations of democracy, summarizing the nature of WPF goals by citing John Dewey, who insisted on the need to cultivate the common good through "the habit of amicable cooperation." Dallmayr again references Dewey when pointing to the larger faith in peace that grounds WPF efforts to nurture trans-civilizational dialogues. He quotes Dewey's statement that a "genuinely democratic faith in peace is faith in the possibility of conducting disputes, controversies and conflicts as cooperative undertakings in which both parties learn by giving the other a chance to express itself." While Dallmayr remarks that Gandhi or King could have expressed this same sentiment, Zapatero himself exemplified a standard of faith equivalent to Dewey's, in his remarks to the 59th General Assembly session on the possibilities for hope, change, and collective action in the face of terrorism and world calamities.

Referring to the Spanish withdrawal of troops from Iraq, Zapatero affirmed at that session, "Peace must be our endeavor. An endeavor that requires more courage, more determination and more heroism than the war itself." He offered the following proposal for an *Alliance of Civilizations* that would serve as just such a peace endeavour, addressing the challenges of our day:

> Thus, in my capacity as representative of a country created and enriched by diverse cultures, before this Assembly I want to propose an Alliance

of Civilizations between the Western and the Arab and Muslim worlds. Some years ago a wall collapsed. We must now prevent hatred and incomprehension from building a new wall. Spain wants to submit to the Secretary General, whose work at the head of this organization we firmly support, the possibility of establishing a High Level Group to push forward this initiative.

Zapatero's call for an *Alliance of Civilizations* to address the Southwest Asia tragedy and the larger problems of war and extremism not only rose from his familiarity with the discourse on democracy and dialogue that was made prevalent with the establishment of the World Public Forum, but it was also intimately connected to his awareness of Spain's cultural heritage of peace and prosperity, a heritage that has grown through the dialogical cooperation of the country's Christian, Muslim, and Jewish populations. In an article written by Cynthia Rush for Executive Intelligence Review it is pointed out that Zapatero's insistence on a positive view of his nation's history of collaboration contrasted with his predecessor's pessimistic view of past culture and human relations in Spain, and his defensive call for "a holy Crusade" that would vindicate not only the recent Madrid bombing attributed to al-Qaeda but also ancient, 8th-century violations of Spain by the Moors. Speaking on the same day as Zapatero addressed the General Assembly, Jose Maria Aznar, the former Prime Minister of Spain, offering an inaugural lecture as a visiting professor at Georgetown University in Washington, D.C., adopted the logic of "the clash of civilizations," proudly declaring in his speech that just as the Catholic Monarchs, Ferdinand and Isabella, had led the "Reconquista" at the end of the 15th century, so too would Spain now refuse to surrender its identity, "to become just another piece of the Islamic world." To top off this triumphal claim, he then added, somewhat threateningly, that "many radical Muslims remember that defeat. Osama bin Laden is one of them."

Aznar's stance, which provided dramatic contrast to Zapatero's, reflected ideology and a mode of thinking that many recognized as having been similarly employed by the Bush administration to justify the invasion of Iraq. Such a stance was directly related to an argument put forth by Samuel Huntington in his *The Clash of Civilizations and the Remaking of World Order* (Huntington 1996). But in this place of Clash of Civilizations, Zapareto addressed with a bold spirit of heroism when he proposed the *Alliance of Civilizations* and reminded those gathered at that 59th session:

> The United Nations was born out of necessity and ideals. It was built by women and men who asserted their faith in the understanding among peoples and cultures. They left us Utopia as a legacy. They thought that every goal was within their reach: settling old conflicts, eradicating poverty, ensuring rights for every human being; and today we could ask ourselves: what is within our reach?
>
> Almost everything. True, human kind's history does not provide us with many reasons to be optimistic. Nor does the world of today offer

us many reasons to feel superior to the women and men that preceded us. . . . No, we human beings cannot feel very proud of ourselves today. We definitely need to change this. We, the Spanish women and men of today, are resolved to make it possible for those who come after us to be able to say: "Yes, they did it!"

Zapatero's proposed initiative received co-sponsorship from the governments of Spain and Turkey under the leadership of Prime Minister Erdogan. Secretary-General Kofi Annan launched the initiative in July 2005. In a statement issued on July 14, 2005 by the Spokesman for Annan, the newly formed Alliance was described as an effort to address the "hostile perceptions" that "foment violence" and to bring about the conditions that are conducive to healing divides.

UNAOC began with the initiative of Zapareto who wanted to break away from American hegemony and withdraw from Spanish participation in the Iraq War. Spain and Turkey came together to found this Alliance. As two observers write:

> Spain has redirected its external priorities toward the Mediterranean in a broad sense (the Euro-Mediterranean Process, the Magreb countries, the Middle East etc.). What is more Spain considers that the Mediterranean region and the relations between the North and the South of Mediterranean can be best understood as the best example and a test case of Alliance of Civilizations.
>
> (Mestres and Lecha 2006: 121)

As requested by Zapatero in that September 2004 address, Kofi Annan assembled a High-Level Group (HLG) consisting of 20 individuals from a range of backgrounds, including policy-making, academia, civil society, religious leadership, and the media. It was co-chaired by Frederico Mayor of Spain and Mehmet Aydiin of Turkey. All religions and civilizations were believed to be represented. Among the members of the HLG was former Iranian President Mohammad Khatami, who had earlier proposed the Dialogue Among Civilizations initiative discussed earlier. Khatami offered the group a particularly creative, poetic, and broad spiritual perspective. In a speech before the United Nations General Assembly on the United Nations Year of Dialogue among Civilizations, expressing his guiding insights about the underlying causes of war, he stated:

> Should human beings be deprived of compassion, and be divested of morality, religious spirituality, sense of aesthetics, and the ability to engage in poetic visualization, and should they be incapable of experiencing death and destruction through artistic creativity, then horrendous hidden forces of the unconscious should wreak havoc, death and devastation upon the world of humanity.

Additional HLG members included Archbishop Deshmond Tutu, South African Nobel laureate, Professor Pan Guang, highly recognized for his outstanding contributions to China-Russia Relations, and Arthur Schneier, recipient of the Presidential Citizens Medal for founding the Appeal of Conscience Foundation. In the year spanning from November 2005 to November 2006, the HLG met five times, in Spain, Doha, Qatar, Dakar, Senegal, and finally in Istanbul, where members presented the final report to Kofi Annan and to Prime Ministers José Luis Rodríguez Zapatero and Recep Tayyip Erdoğan. Secretary-General Kofi Annan had outlined his vision for the Alliance of Civilizations during early meetings of the High-Level Group. He directed members to focus on the initiative as a response "to the need for a committed effort by the international community—both at the institutional and civil society levels—to bridge divides and overcome prejudice."

The High-Level Group published a report which emphasized the need for dialogues and celebrated diverse cultural identities as a source of human prosperity and emphasized the role of traditions and customs in developing and transmitting modern identity. It also considered group identities in Latin America, Africa, and Asia that have been challenged in a "globalized" world due to market-based forces of commodification and homogenization and problems related to urbanization, such as dislocation, loss of traditional lifestyles, and environmental degradation. Most significantly in relation to its larger declared mission, the report sought to address the emergence of extremism by distinguishing it from fundamentalism and by considering the problems both of wide-scale modern secularization and of exploitation of religion by ideologues.

The efforts of the Alliance of Civilizations can be understood to focus primarily on healing relations between Western and Muslim societies, by clarifying the true roots of present conflict. While propagandized versions of "ancient history" are used to promote a "clash of civilizations" mentality that enforces patterns of conflict, the Alliance of Civilizations diagnoses the actual roots of modern-day hostilities in the more basic and reactionary dynamics that took place in the 19th and 20th centuries, "beginning with European imperialism, the resulting emergence of anti-colonial movements, and the legacy of the confrontations between them."

In this regard, the Alliance represents a fresh kind of historical perspective that awakens consciousness about a central dilemma of the 21st century: how to restore and protect processes of humanization in a global world that has become modernized in large part through political, economic, and cultural forces of domination and oppression that ironically purport to advance human progress. Major historical narratives that the Alliance hopes to heal and transform through new forms of perspective-taking and policy recommendations are the partitioning of Palestine by the UN in 1947, viewed as a form of colonialism by many in the Muslim world; the 1953 coup in Iran, which related to Cold War powers vying for influence in strategic resource rich regions and which seriously backfired in the aftermath of this coup; the

1979 Soviet invasion and occupation of Afghanistan and the US arming of Afghan resistance to it, which later led to increased animosity against the West when the Taliban regime gained control of the country; the 2001 Al Qaeda terrorist attacks on the USA, which then provoked a forceful retaliation against the Taliban regime in Afghanistan that was used as one of the justifications for the invasion of Iraq. This last narrative has in particular fed the perception among those in Muslim society that the West is unjustly aggressive towards them.

Although the Alliance focuses on the Western–Muslim society relationship, and its historical narratives, its overall approach to establishing peace and harmony is considered useful for bridging diverse kinds of cultural divides. Its list of general policy recommendations are firmly in line with its guiding principles, and consist of: a renewed commitment to multi-lateralism; a full and consistent respect for international law and human rights; coordinated migration policies consistent with human rights standards; combating poverty and economic inequities; protection of the freedom of worship; exercising responsible leadership; the central importance of civil society activism; and establishing partnerships to advance an Alliance of Civilizations.

The programmes outlined in the second part of the HLG report, or Implementation Plan, relate to four key areas: youth, education, media, and migration. A major aim of the youth effort is to address issues of socio-economic and cultural alienation and to expand opportunities for youth mobilization. This includes student exchange programmes, sports activities, and political involvement that provide new opportunities for promoting cross-cultural understanding and respect for diversity. The Alliance emphasizes that it views youth not only as a source of mobilization but also as autonomous actors and partners. The Youth Solidarity Fund (YSF), the Alliance's Fellowship Programme, and the Global Youth Movement are examples of their initiatives.

The planning for new initiatives in education to support Alliance goals involves civic peace education, global and cross-cultural education, exchange programmes, particularly at the post-graduate level and in the sciences, media literacy education, and education on religion. Examples of programmes developed by the group are The Alliance of Civilizations Summer School and The Alliance of Civilizations Education about Religions and Beliefs Online Learning Community.

In terms of media, the Alliance analyses how the media shapes our views and focuses on "amplifying the role of media in furthering public understanding of cross-cultural issues." One major effort has been the creation of Global Experts, described as "an online database of nearly 400 experts who provide free analysis to journalists on cross-cultural news stories." It is supported by a network of partners from the media, academia, civil society, and the international community, including the European Commission, the International Center for Journalists, Search for Common Ground, International Crisis Group, and the Global Forum for Media Development. The experts on the website provide quick reactions, give quotes, and share their views with media professionals in response to breaking news events. Global Experts is additionally described as supporting "the work of journalists, editors and producers, particularly but not exclusively in times of cross-cultural crises." It encourages a broadened choice in available commentaries by nurturing a diversity of experts and opinion leaders to share their perspectives on issues that "go to the core of relations among diverse communities." Media literacy is also an area of focus as is press freedom and responsibility and the impact of entertainment media on social perceptions, particularly in representations of ordinary Muslims in Western mass media.

Finally, the Alliance's programme in migration focuses on the dynamic dimension of migration, proactive strategies for addressing migration, benefits and challenges of migrations, combating discrimination, supporting expanded inclusive dialogue at all levels, and political, civil society, and religious leadership in the West that can help set the tone for debates about immigration. In partnership with the Media Programme, the Migration Programme has also held a number of trainings for journalists on covering migration. The Alliance has also developed PLURAL+, a youth video festival on migration, diversity, and social inclusion, and the website "Integration: building inclusive societies" which has aimed to provide information on good practices of integration of migrant populations and highlight successful models of integration to counter stereotypes.

Jeffrey Haynes, a political scientist from London Metropolitan University, writes about the scope of the Alliance of Civilizations and its impact worldwide. In his work, he elucidates eight main projects that were created to empower inter civilizational dialogue with a focus on education, youth, migration, and media. These are (a) summer schools for 75–100 students aged 18–35 years focused on "fostering diversity and global citizenship, reducing stereotypes and identity-based tensions, promoting intercultural harmony and social justice" in relation with EF Education First, a private international education company; (b) the "Intercultural Innovation Award" and (c) "Intercultural Leaders Award" run in partnership with the German luxury car business, BMW Group; (d) initiatives catered to further "Media and Information Literacy"; (e) a video festival named "Plural + Youth" in relation to the International Migration Organization (IMO); (f) the annual "Hate Speech Conference" aimed at tackling narratives of hate in the media; (g) a fellowship programme uniting youth from the Western and Muslim worlds; and finally (h) Youth Solidarity Fund that supports youth-led endeavours that unite people from diverse backgrounds (Haynes 2017: 1130).

In his valuable analysis, Haynes points out that while these efforts are tailored to institutionalize cross-cultural dialogue, it is hard to gauge the impact of these special projects in the realm of inter-relations and global justice. This is because those who benefit from the projects are not only fewer in number but also from the already existing elite (ibid). Since its inception a decade ago, he elucidates that there has been some slow progress and significant drawbacks. The UNAOC has fallen short of its planned ability to regularly bring together governments, International Organizations (IOs), and Civil Society Organizations (CSOs). Additionally, although the UNAOC is established as an entity of the United Nations (UN), its financial, organizational, and policy-related weaknesses hinder its ability to advance global justice through relations between the West and the Muslim world (ibid: 1137).

Furthermore, Haynes pays special attention to the interfaith work that is carried out by the Alliance of Civilizations. The Alliance also works with The Committee of Religious NGOs in the UN which is a "coalition of 38 Faith Based Organizations (FBOs) that seek to serve as a coordinating function for non-state faith-based entities at the UN, including in relation to UNAOC" (Haynes 2018a: 53). They also collaborate with Religions for Peace, and United Religions Initiative (URI). However, he brings to light two major drawbacks. The outreach is restricted to elite groups and there is minimal effect on reducing religious-based conflicts. Haynes stressed that "religious leaders: such as, women, indigenous leaders, younger peer leaders, and ordinary people of all identities" who often lead and empower community cooperation are not recognized as they do not hold any "official" titles (ibid).[5] Similarly, through a survey of three major religious NGOs – The Committee of Religious NGOs at the UN, Religions for Peace, and United Religions Initiative – Haynes found that it was impossible to conclude a definitive judgement of the activities of FBOs. The activities of the 350+ FBOs were "too disparate and too undocumented." Haynes stressed that interfaith dialogue at the international level is well meaning but its effect on reducing conflicts and tensions was unclear (Haynes 2018a: 53).

An important critique from the Report according to Haynes was "a lack of global justice as a key reason for increased friction and conflict between the West and the Muslim world" (Haynes 2017: 112). But Haynes insightfully argues how this also raised the issue of plural conceptions and realizations of justice and the need and responsibility to listen to Islamic narratives. He says:

> Note however that the conception of justice in Islam is a much more profound and overarching value compared to that found in the UN system. This is because Muslims believe that the conception of justice in Islam comes not from humans but from a higher source: God.
>
> (Haynes 2017: 1132)

He supports this by referring to the work of Islamic scholar and legal philosopher, Abduallahi An-Naim who preaches the need for a "people-centred" approach to rights and global justice which would in turn empower global actors and remove reliance on state regulation (ibid: 1134).

Haynes (2018b) offers a deeper analysis of the working and members of the UNAOC in his book *The United Nations Alliance of Civilizations (UNAOC) and the Pursuit of Global Justice: Overcoming Western Versus Muslim Conflict and the Creation of a Just World Order*. Haynes tells us that UNAOC "is the only public entity publicly concerned with non-secular understandings of the world." Since its creation, UNAOC formed a Group of Friends consisting of both UN Member Nations and other international organizations such as the Vienna Based King Abdullah bin Abdullaziz International Center for Interreligious and Intercultural Dialogue (KAICIID). By 2009, there were about 100 members in the group. However, Haynes notes that many member states of the UN have still not joined the alliance as they consider it only a talking shop and "ineffectual" (Haynes 2018b: 107). Moreover, UNAOC has close relationships with governments of three countries—Saudi Arabia, Azerbaijan, and Kazakhstan—while dealing with challenges like terrorism. Ironically, these nations that fund the UNAOC themselves curb religious freedom in their own nations. Haynes recounts his personal experience at the meet at Baku where the leaders of both Azerbaijan and Turkey used the forum to express their anti-Armenian biases, which was one of the reasons why then U.N. Secretary General Ban-Ki Moon avoided attending the meet.

Second, although the Report recommended the creation of international, national, and local councils to carry out the work, no local council has been formed (ibid: 144). Furthermore, "Only 9 of the 28 EU member states (32%) drew up national strategies (Bulgaria, Croatia, Denmark, Italy, Malta, Portugal, Romania, Slovenia, Spain)" (Haynes 2018b: 89–90). Finally, he highlights the perspective of senior representatives of a Western international organization who view that the UNAOC is "not very good at consistently pursuing its objectives." He reasons that the UNAOC is perceived as a "soft power" tool to address friction between Western and Muslim societies. While in a post 9/11 and 11-M world which has experienced the rise of Islamic State, Boko Haram, and Al Shabaab, many UN member states are more convinced of the effectiveness of "hard power" for there is no evidence that Alliance style dialogue would impact the actions of jihadi groups.

In his 2018 paper, published by the London School of Economics, Haynes questions the core mission of the Alliance and the impact of its dialogues while keeping in mind its shortcomings over the years. According to him, the Alliance sets itself apart from the hard power of military and economic clout as a soft power striving to common civilizational ground against extremism and terrorism. He clarifies that this does not signify complete, doubt-free agreement between the leaders of the Christian and Muslim world but rather an effort to establish a shared set of values in accordance with UN's foundational document: the UN Charter of Human Rights. He finally adds that the Alliance has the capability to save those at the brink of civilizational enmity if they work flexibly with significant stakeholders while overcoming three main obstacles. First, there exists a disagreement on the intercivilizational goals and means to measure it. Second is a lack of support from the UN, especially of financial viability and, finally, 15 years post 9/11, noting a shift to more topical concerns such as fixing the economy or regional wars, such as Syria, rather than improved intercivilizational relations (Haynes 2016).

In a recent reflection on this theme, Haynes (2021) raises some foundational questions to some of the assumptions of UNAOC. UNAOC wants to create more understanding and dialogues between the West and Muslim countries which is based upon an assumption that there is inherent conflict between them. Haynes says that in his field work with Muslim countries, his interlocutors told him that they do not perceive the relationship between the West and the Muslim world in such an essentialistic antagonistic way. UNAOC is therefore based upon assumptions, fear, and anxiety of the West and it is not based upon actual needs, perceptions, and aspirations of the Muslim world.

Another important arena of cross-civilizational dialogue today is a global goal like United Nations Sustainability Development Goals. From Millennium Development Goals, the UN has moved to Sustainability Development Goals. But each of these goals needs to be open to cross-cultural and cross-civilizational dialogues on the meaning and process of realization of these goals. Take, for example, the SDG goal about poverty. SDG goal number 1: No Poverty invites us to engage in cross-civilizational and inter-civilizational dialogue about the meaning and realization of no poverty. For example, it invites us to understand the distinction between voluntary poverty and involuntary poverty. No poverty does not mean that, in our own lives and society, we should live with less resources and consumption and with voluntary poverty as envisioned by individuals and movements such as Jesus, St. Francis, Gandhi, Pope Francis, and the degrowth and simple living movements around the world. No poverty does not necessarily mean war on poverty or eliminate poverty (Garibi Hatao) as it was valourized during the time of President of Lydon Johnson of the USA and Prime Minister Indira Gandhi. These slogans and discourses led not so much to transformation of conditions of poverty but their stigmatization and sometimes lowering of their self-esteem and dignity. SDS goal of No Poverty invites us to realize how we can live with voluntary poverty while transforming conditions of structural and cultural poverty.[6]

Dialogue of Civilizations Research Institute

Dialogue of Civilizations Research (DOC) Research Institute is a private think tank based in Berlin and Moscow with a new office at New Delhi, India. It has also been working on cultivating dialogues across cultures and civilizations. It has nurtured Rhodes Forum where politicians, media persons, and thinkers from both the West and the East meet. Dr. Vladimir Yakunin is a co-founder of the Forum and presently Chairman of the Supervisory Board of DOC Research Institute. In an interview with him published in Helsinki Times, Yakunin tells us:

> Fundamentally, the aim of the forum is to look at latent or existing conflicts, help to understand the root causes, and to try to arrive at practical

solutions. Most international forums are guilty of trying to look at problems through a single lens. At DOC, we are trying to avoid that mistake by working hard to bring in people from all countries, cultures and perspectives. This year we had delegations from Iran and Israel. In the last we have had Sunnis and Shias at the same table. This makes for some very *hot* dialogue at times, but it's better to have hot dialogue than hot war.

Dialogue of Humanity

Dialogue of Humanity is an initiative started by Genevie Ancel and her friends in Lyons, France. I had taken part in their meet in July 2010. It was held in the big park in the city. There were different sessions and workshops spread all around the park, and people from many different countries and different towns and cities had taken part in this. There were sessions offering massage, yoga, and contemporary important global themes such as social movements and women's question. The park turned into a festive place of learning and dialogue. There were sessions on body movements and creative dance. I attended one such session offered by Regis. Regis has experience in creative dancing and he also had experience in Indian spiritual tradition as he spent some time with a spiritual teacher in India. Many of workshop organizers had some cross-cultural experience. After the dialogue, I visited some of the participants. I visited Regis in his home up in the mountains. I also visited Marie-Joe who lives in the Pyrenees—a mountain chain in France. Both of them have had experience with other cultures. Marie-Joe, for example, has been walking with Sri Aurobindo and Mother from India.

As has been mentioned before, in 2010, Dialogue of Humanity was accompanied by a walk called Poverty Walk in which some of the local activists with links to Ekta Parishad had taken part.

Dialogue of Humanity has now become a global phenomenon. It takes place in Fireflies, Bangalore, India, and in other countries such as Brazil. In Brazil, Emerson Andrade Sales and Deborah Nunnes teaching at the Federal University of Bahia had taken the initiative to organize Dialogue of Humanity in Salvadore, Brazil. Both of them also take part in Dialogue of Humanity meeting in Fireflies. I recently met with Emerson and Deborah in Fireflies Dialogue of Humanity meeting in March 2022 and had a discussion with Emerson about the Dialogue of Humanity meeting in Brazil. Emerson told me that he along with the help of some of his students in the University had started this. It then attracted a lot of attention and participation from the people in the city of Salvadore and the surrounding areas. It became a space of dialogue and alternative explorations. But the work load became too much on him and he could not continue the Dialogue subsequently.

I had also met with Emerson and Deborah in earlier Fireflies Dialogue of Humanity as I have taken part in several of these meetings since 2010. It brings activists and scholars from different fields and countries. For example, in 2011, the theme of the Dialogue was on Food Sovereignty. There was a discussion on reviving millet cultivation which would also contribute to climate change mitigation. There was a millet festival. There were many dishes prepared, such as millet dosa.

Fireflies has established itself as an intercultural centre. It has a series of publications called Meeting Rivers which brings together dialogical reflections on many contemporary issues. Fireflies strives to be a place crossing over many borders. In its campus, it has many pictures of goddesses and Divine Mothers. Its inter-religious prayer hall has symbols of spiritual seekers from many traditions such as Sri Ramakrishna and Mother Teresa.

Center for Peace and Spirituality International and the Work of Maulana Wahiduddin Khan

Maulana Wahiduddin Khan is an important Islamic thinker in contemporary India. He has established Center for Peace and Spirituality as a centre of peace building and dialogue. For his many-sided contributions, Khan has been bestowed with Padma Vibhushan, India's third highest civilian award in 2021. In his Center's website, the following is written about him:

> In 1992, when the atmosphere was so highly charged throughout India due to the Babri Mosque incident, he felt the necessity to convince people of the need to restore peace and amity between the two communities, so that the country might once again tread the path of progress. To fulfill this end, he went on a 15-day Shanti Yatra (peace march) through Maharashtra along with Acharya Muni Sushil Kumar and Swami Chidanand, addressing large groups of people at 35 different places on the way from Mumbai to Nagpur. This Shanti Yatra contributed greatly to the return of peace in the country.

Khan was born in 1925 and had taken part in the freedom struggle of India along with Gandhi. He studied in a Madrasa, an Islamic seminary, in Azamgarh. But after this, he realized the value of learning English and science—the topics and themes on which he did not have training. He then studied these on his own. He then tried to make a dialogue between science and religion. He strove to revive true religion from its original sources and presented it in the modern idiom. During our conversation in Delhi in June 2015 Khan told me that once he was visiting Cordoba, the seat of Islamic civilization in Andulasia. Many Muslims lamented that it was such a peak of high achievement of Islamic civilization. But then Khan felt:

> What is the point in lamenting about our achievements in the past? We should go ahead with our efforts to advance and progress in the present. We should go ahead with modern science and technology. This is also the gift of Allah. Muslims should learn science and technology and be part of the modern world.

Khan brings a spirit of science to his study of Quran and to his religious and spiritual journey. While appreciating the work of the Sufis in creating peace and harmony in the world he feels that Sufis may at times promote a kind of emotional outburst which is not accompanied by deeper works of reason.

During our conversation, we discussed about reading Quran and the possibility of asking critical questions. We also talked about dialogue and peace. Khan felt that most of the time, dialogue becomes reduced to only dialogue and it does not become anything more than a talking shop. About peace, he said: Quran says that peace is the best. Quran does not say that war is the best.

Khan and his centre, Center for Peace and Spirituality, have been spreading the message of peaceful Islam. He edits a journal named *Al Risala* in Urdu and *Spirit of Islam* in English which has a wide following. His centre also works in Kashmir. During our interview, Khan mentioned that many Kashmiri youths have shun the path of violence by reading his work and the works of his centre.

In his book, *The Prophet of Peace*, Khan (2009) challenges us to understand how Islam and the Prophet of Muhammad wanted to avoid war and use it only as a last resort. For dialogue, Khan pleads for development for creative and critical thinking. In his book *Leading a Spiritual Life*, Maulana Wahiduddin Khan (2016) tells us:

> Critical thinking is constructive thinking. It is a phenomenon of creativity. And creative thinking is the only way to intellectual progress. Creative thinkers are always able to discover new things, but creative thinking cannot be developed without critical thinking. You should, therefore accept criticism with a tranquil mind and you will soon discover that your critic was an intellectual enabler.
>
> The greatest weakness of every man or woman is that he or she lives in the boundaries of his or her own mind. This kind of thinking tends to induce self-conditioning, which is not always a positive nature. No one, except your critic, can de-condition your mind.
>
> (Khan 2016: 218)

Sonam Kalra's Sufi Gospel Music Project

Sonam Kalra is a creative musician who in his project brings many different traditions of religion music especially Christian Gospel music and Sufi music together. The projects like Kalra's are very important parts of contemporary landscapes of dialogues.

Roots[7]

Roots is an initiative in dialogue and peace-building centred around a piece of land in the Gush Etzioni-Bethelhem area in West Bank which brings Israelis and Palestinians as well as concerned people struggling for peace from other parts of the world such as the USA together. It was founded in 2014 by Ali Abu Awwad and Saul Judelman after the meeting of Israeli and Palestinian activists on a strip of land owned by Ali's family. A Rabbi, Rabbi Hanan Schlesinger, soon joined this effort and it also quickly received support from John Moyle, a pastor of Mission and Social Justice from the USA. Roots works with individuals, families, and communities "at the heart of the conflict" trying to shift "hatred and suspicion towards trust, empathy, and mutual support."

Roots has several projects in mutual learning, trust-building, and peace. The following are some of these:

> Informal Actions in Helping People: Here Roots helps people in need and arranges hospital visit and visit of civil and military administration to help people
>
> Local Leaders Programme: These local leaders from Palestinian and Jewish backgrounds together. Roots has engaged 55 local civil and religious leaders. These leaders work on "tolerance education in local schools, medical coordination and programmes which promote leadership against violence."
>
> Summer Camps: Roots organizes summer camps for Palestinian and Jewish students where they stay together
>
> After school programme: It hosts "photography workshops which bring Palestinian and Israeli children together in ongoing meetings throughout the year."
>
> Pre-Army Academies: This works with Israeli young people going to join the Army to be more sensitive to the conditions of Palestinians. It organizes two months exposure to Palestinian narratives.
>
> Responding to Violence, Points of Calms

In a situation of heightened violence, Roots facilitates communication channels between neighbouring Israeli and Palestinian local leaders to reassure each other and be a source of calm and peace rather than immediately react with violence. Roots trains people on how to respond to violence in a non-violent manner and remain calm in front of provocation.

Dual Narrative Programme

Living with each other's stories is the foundation of our shared life and this dual narrative programme of Roots brings Palestinians and Jews for five months to explore "personal identity, connection to the land and each other's narratives." There is also an interlinked programme of family meetings where both Jews and Palestinians spend time with each other's families and try to get new strength for a new dream and steps to peace partaking in the hospitality and sharing of bread in the House of Abraham. Roots also organizes inter-religious exchange and learning of language.

Roots has been promoting a culture a non-violence, trust-building, and responsibility in a difficult environment of hostility and escalation of violence.

It has been urging both Palestinians and Jews not be locked in their prisons of victimhood and take responsibility for peace and co-existence in their lives. Rabbi Hanna Schlesinger is one of the main activists of Roots, and in an article, he challenges all concerned thus:

> But there is also the culture of victimhood, the sense that it is all the fault of the other side and it is only the other side's responsibility to fix this mess. Too many Palestinians are waiting for the other side to take the initiative. I get that as well; many Israelis feel the same way. But as long as both sides continue to see themselves only as victims of the other, we will remain mired in this conflict for decades to come. Palestinians, just like Israelis, must take responsibility.[8]

Ali Abu Awwad and Rabbi Hannan Schlesinger tour around the world jointly speaking of their approach to peace. They were speaking in Denver, Colorado, in a synagogue in October 2015. I had listened to them. Their joint sharing created a lively interaction and interest in the audience.

Foundation for Universal Responsibility of His Holiness the Dalai Lama (FURHHDL)

His Holiness Dalai Lama has been striving for peace and dialogue and has been an important presence in our world today. The Dalai Lama also urges all of us to realize our own inter-linked paths of responsibility. Out of the Nobel Prize money that the Dalai Lama received, he has created Foundation for Universal Responsibility. The Dalai Lama says that the Foundation wants to "implement projects to benefit people everywhere, focusing especially on assisting nonviolent methods, on improving communication between religion and science, on securing human rights and democratic freedoms, and on conserving and restoring our precious Mother Earth."[9] The Foundation has four programmes: inter-faith council, humanism and social awareness, Peace Building, Non-Violence and Co-Existence, and Ethics in Education. All these programmes focus on philosophies of compassion and contribute to both proper cognitive development and spiritual development of individuals and societies. FURHHDL organizes dialogues among religious leaders and also organizes Youth Inter-Faith Pilgrimage which brings youth to places of different religious pilgrimages.

The Foundation also works with the following Affiliating Organizations. The Dalai Lama Foundation is based in Silicon Valley and it is an international organization with chapters in different countries. Its main programme focus is in the area of education and strives to develop and deliver curricula for ethics and peace. FURHHDL's other affiliating organization is The Dalai Lama Center for Peace and Education which is based in Vancouver. It works on realizing Dalai Lama's lifelong "commitment to compassion and inner well-being." It also works on education of heart and has a project on "Educating the Heart: A Global Aspiration." It works on both cognitive development and spiritual self-development of individuals and societies. The Foundation is also affiliated with the Dalai Lama Center for Ethics and Transformation of Human Values at MIT. It is also affiliated with Center for Compassion and Altruism Research and Education at Stanford University.[10]

International Charter of Human Responsibility and Asian Charter of Responsibility

United Nations had promulgated Universal Charter of Human Rights in 1945. Many individuals and groups subsequently felt that we also need to have a Universal Charter of Human Responsibility. Charles Leopold Mayer Foundation had played a key role in taking this initiative. In the Roundtable of Epilogues and Reflections, Julie M. Geredien writes about this effort.

The Universal Charter of Responsibilities from the beginning has been part of a global dialogue. For instance, the Foundation has organized series of meetings on the very term of responsibility and has tried to understand its multiple meanings and languages (Sizoo 2000, 2008). In the book that has emerged from this dialogue, *Responsibility and the Cultures of the Worlds*, Sizoo and her co-walkers from around the world have explored different languages, meanings, and pathways of responsibility. In this dialogue, Makarand Paranjape (2008) from India brings the dimension of *dharma* which means righteous conduct to understand responsibility. He argues how "Responsibility is relational, not absolute or relative" (Sizoo 2008: 24). In this dialogue, Te A. C. Royal and Betsan Martin tell us how, in Aotearoa/New Zealand, indigenous term *Kaitiakitanga* refers responsibility both to the natural world and to the relational world. They tell us:

> Whilst the Maori language does not maintain a word that is directly comparable to responsibility, through its significant ethical dimension, kaitiakitanga holds meaning of relevance to responsibility.
>
> The root word of Kaitiakitanga is "tiaki" which means to care for, to foster, to nourish. A "kaitiaki" is a person or some kind of other agent who practices and administers kaitiakitanga, that is the administration of care toward people or toward "treasures" and valuable items such as land. . . . A Kaitiaki is a person, or a group of people, who understand this interconnectedness and are sensitive to ways in which their impacts may flow throughout a system.
>
> (Royal and Martin 2008: 54–55)

This global dialogue has also led to some critical reflections such as that of Clifford G. Christians who argues, "Responsibility in the United States has the double meaning of accountability and being-in-charge" (Christians 2008: 65). For Christians, responsibility language "has no natural home in the U.S. thinking and policy within the overwhelming regime of negative freedom" (ibid: 70). At the same time, for Christians, there are constructive movements of accountability in American histories and societies such as the Jane Adams Hull Full House (and to this, we can add varieties of social, political, and spiritual movements such as anti-slavery movements, civil rights movements, and movements such as Habitat for Humanity which builds houses for low-income families) "which need to be expanded and publicized for responsibility to have resonance in contemporary America" (ibid: 70).

Such critical and creative engagement with responsibility has also given rise to new initiatives in thought and practice. In India, friends working with the Foundation, such as S. Sudha from Bangalore, have been working on creating an Asian Charter of Human Responsibilities. She and friends have also tried to bring the dimension of gender justice, children's issues, and inter-faith dialogue to the work of responsibility. In 2008, Sudha had organized a dialogue on creating An Asian Charter of Human Responsibility in which many thinkers and activists from India and South Asia such as Jeevan Kumar, Makarand Paranjape, Vijaya Pratap, and the late Nalini Swaris had taken part. She has also taken initiative to create a charter of responsibility for administrators and professionals. In 2015, in an international conference on "International Law" in Delhi, she organized a series of sessions on responsibility and international law. Sudha has also translated the Charter of Universal Responsibilities into Kannada.

Parliament of the World's Religions

Parliament of the World's Religions which was organized in 1893 in Chicago was an important event in modern world which opened the doors and windows of Euro-American world to other religions. Swami Vivekananda from India had taken part in this and he used this occasion to present Vedanta as a universal quest for peace. Parliament also brought people from other religions. This Parliament got revived after 100 years in 1993 in Chicago. After this, there were sessions of the Parliament in South Africa and Australia. The last one was held at Salt Lake City, Utah, in October 2015 which brought nearly 10,000 people from different religions and cultures of the world together. It also brought many grassroots inter-faith work from many religious traditions of the world together. The Parliament began with an assembly of women spiritual leaders of the world who were trying to revitalize the Mother Goddess tradition of spirituality in all religious traditions. This happened before the official inauguration of the Parliament. Revitalizing feminine spirituality in all religious traditions as well as in science and life was an overwhelming flow in the Parliament. Another interesting feature of the Parliament was that Imam Malik from Islamic stream was the President of the Parliament who possibly had inspired a greater degree of creative and transformative participation of Islamic religious seekers and leaders in the Parliament who were discussing about issues of terrorism, violence, and peace. There was a greater degree of participation of indigenous religious and spiritual leaders. I had taken part in this Parliament and could feel that with all these interconnected movements in thought, practice, dialogue, and mutual co-presence, Parliament was becoming a space for nurturance of global responsibility and a new imagination for planetary realizations. Here are some of the photographs from the last Parliament of World Religions which reflect different movements and possibilities.

Figure 3.3 The author with a volunteer at the Langar in the Parliament of World's Religions at Salt Lake City, Utah.

Source: Author's Archive

Figure 3.4 Some of the photos from the Parliament of the World's Religions, Salt Lake City, Utah, in October 2015.

Source: Author's Archive

Figure 3.4 (Continued)

These are some of the pictures of movements from the Parliament of Religions in 2015. I had also taken part in the subsequent and last Parliament of Religions meet in Toronto in November 2018. The meet began with the invocation of indigenous spirituality in Canada and North America. It had brought religious and spiritual leaders, artists, and activists from many countries around the world. During our COVID-19 times, 2021 The Parliament of World's Religions took place online from October 17, 2021 to October 18, 2021 on the theme, "Opening Our Hearts to the World:

Figure 3.4 (Continued)

Compassion in Action." I nurtured a Panel here on the theme of "Opening Our Hearts to the World: Cultivating Transformative Faith and a New Ecology of Hope." In our panel, we explored how we can cultivate transformative faith which is different from blind faith, and such a faith helps us in our dialogues among religions, cultures, and selves and creates new ecologies of hope. We explore how hope emerges from dialogues across borders and movements for change.

Figure 3.4 (Continued)

This chapter describes varieties of movements across borders and initiatives in dialogues in our contemporary world. But despite these movements, violence and intolerance continue unabated in this world which calls for deeper transformative dialogues and movements of responsibilization. The recent political movements in the world such as xenophobic nationalism and religious and political majoritarianism have posed new challenges to visions and practices of dialogues.

Notes

1 Ambar Istiyani works with Percik and she writes in her email of June 22, 2020 about the work of Percik:

> During this COVID pandemic, we work from home. Only when it is urgent, we go to the office. There is hardly a meeting for interreligious dialogue. The last Sobat interfaith meeting was last year in Kudus, Central Java where Sedulur Sikep, a Javanese community lives. Sedulur Sikep is a minority group in Indonesia that suffered discrimination in term of civil rights. The government did not recognize their local belief, so that it was not easy for them to get administrative documents for instance, like marriage or birth certificates. So, we discussed and shared about the challenges of Sedulur Sikep in dealing with changing times (political, social, economic, etc.), including their education (especially local belief subjects in schools).
>
> Even so, we hold some online discussions recently to respond to the pandemic effects. As you may know, Sobat interfaith forum has some derivative programs like interfaith forums for children (Sobat Anak), women (Kata Hawa), and young people (Sobat Muda). Sobat Anak facilitators and some teachers of religious based institutions held an online meeting to share about the challenges and strategies in facilitating children during this pandemic. Kata Hawa held a meeting to discuss how to handle children studying at home. Sobat Muda gave some food to those who are in unemployment because of this pandemic. Additionally, we also still facilitate some couples to hold their interfaith marriage.

2 Photo and video of my dancing Aum Salam Walakum.
3 Khatami et al. 2001, "Dialogue among Civilizations," retrieved from the web.
4 This section on Alliance of Civilizations builds upon a text originally written by Julie M. Geredien, a friend of the project, and subsequently updated. Julie has also joined us in our book with her thoughts on International Charter of Human Responsibility in the Afterword conversations on this book.
5 United Religions bring this critique to the fore and "The URI perspective is that while official religious leaders are honoured as wise partners in interfaith dialogue and cooperation, so too are 'ordinary' people, also working to achieve interreligious and intercultural harmony" (Haynes 2018a: 57).
6 On April 29, 2017, Professor Jeffrey Sachs, a noted economist and now Director of Earth Institute of Columbia University, was speaking on the occasion of the bicentennial celebration (April 27–29, 2017) of Harvard Divinity School. In his address, Professor Sachs also spoke about the need to fight poverty. During discussion, I raised the issue of limits to discourses such as war on poverty which forgets such ways of transforming poverty such as simple living and voluntary poverty. By embracing voluntary poverty and simple living, we also contribute to reduction of suffering and structural poverty in the world. After the question answer, one Christina nun came up to me and said how important it is not to forget such challenges of simple living and voluntary poverty on such matters.
7 The narration presented here mainly builds upon materials on Roots available from its website and some other internet resources.
8 Rabbi Hannan Schlesinger writes about this in his blog, "Show Me Your Face" retrieved from the internet. The full text deserves our careful consideration:

> I am what you might call an alternative peace activist. The return to the biblical heartland of Israel—Judea and Samaria (the West Bank)—where Abraham spoke with God and where King David ruled, is a value that animates my being. This modern-day miracle is the fulfillment of the dreams

of our forefathers. But I also hear the echoes of the biblical prophets in my ear, crying out for righteousness and justice for all. I cannot abide the lack of rights and the loss of dignity that the Palestinians suffer daily. I witness it directly and hear about it in the first person, and it tears me up inside.

I identity with the right, and I identity with the left. You cannot easily label me.

And it is not only me. Hundreds of settlers in the Gush Etzion area have gathered around our new movement: Roots/Shorashim/Judur—a local Israeli-Palestinian initiative for understanding, nonviolence and transformation. We are challenging people to remove the blinders from their eyes, to see the other side and recognize his humanity and his story. Without abandoning our own truth, we are learning to see the other's truth as well.

Many Palestinians have also joined our efforts. They also are learning to change their perception of the other and to recognize our truth. But there are not enough of them. A tipping point seems almost to have been reached. We are on the verge of recruiting thousands of settlers for our dialogue activities, but it is not clear if we will have enough Palestinians for them to talk to.

There are many reasons for this. So many Palestinians are in dire economic straits. They cannot find the time or strength to engage in dialogue while they are barely able to put food on the table for their families. I get that.

But there is also the culture of victimhood, the sense that it is all the fault of the other side and it is only the other side's responsibility to fix this mess. Too many Palestinians are waiting for the other side to take the initiative. I get that as well; many Israelis feel the same way. But as long as both sides continue to see themselves only as victims of the other, we will remain mired in this conflict for decades to come. Palestinians, just like Israelis, must take responsibility.

(retrieved the website of Roots)

9 Retrieved from the website of the Foundation.
10 In a talk on Human Responsibility in 2008, the Dalai Lama says:

That is the reality. According to that reality, our centuries-old concept that "us" and "them" are independent is, I think, outdated. Now, particularly in these modern times, with the economic conditions, the environmental issues and the sheer size of the population, everything is interdependent. So, in those circumstances, a Buddhist concept is that you should consider all sentient beings as the mother sentient being, to whom you should develop the same sense of closeness as to your own mother. So, according to theological religion, all creation is created by God. So, we human beings, other sentient beings and the whole world were created by God. A Muslim friend told me that a true Muslim should love the whole of creation as much as they love God. So different words, a different approach, but the same meaning. Therefore, there is the idea that there is a sense of global responsibility, that we should develop a sense of concern for the whole of humanity, the whole world. That eventually develops. For more than 30 years that has been my concept, and it is still relevant. More and more people seem to agree with it.

4 The Calling of Socio-Spiritual Responsibilization

Corporate Social Responsibility, Climate Change, Anthropocene, COVID-19, and Beyond

> In the Anthropocene era the dominant insight is a renewed realization that the fate of the human species is bound up with a collaborative relationship between humans and nature, rather than the earlier beliefs that the former could easily dominate and exploit the latter without serious adverse consequences.
>
> Richard Falk (2016), "Introduction," *Exploring the Anthropocene: Towards the Year 2020*, p. 4

> Corporate Socio-Spiritual Responsibility (CSSR) implies the collective social responsibility of all spiritually inclined individuals, institutions, organizations, Corporate Business Houses and Corporate Religious Houses/Congregations in the world to work together to promote a culture of peace and sustainable development on earth based on a unitive spiritual consciousness. Peace and sustainable development in the world are impossible without inter-religious harmony and co-operation. Hence, Corporate Socio-Spiritual Responsibility also implies inter-religious harmony and co-operation for a culture of peace and sustainable development on earth based upon a unitive spiritual consciousness.
>
> —Acharya Sachidananda Bharati (2017), *The Air Plot: Socio-Spiritual Foundations of Integral Revolution*, p. 123.

Introduction and Invitation

In this chapter, we discuss visions and practices of socio-spiritual responsibilization in fields such as corporations, climate change, Anthropocene, COVID-19, and related fields. We begin with Corporate Social Responsibility (CSR) and then discuss corporate spiritual responsibility and then other issues such as climate change, Anthropocene, COVID-19, and responsibility.

Corporate Social Responsibility

Corporations exist in society, but they do not always realize their responsibility to self, culture, societies, and the world. Corporations are conceptualized as legal persons with rights, but unlike human beings, they cannot be

DOI: 10.4324/9780429347481-4

held culpable for their wrongs. As Edward Thurlow, First Baron Thurlow (1731–1806), Lord Chancellor during the impeachment of Warren Hastings, wrote: "Corporations have neither bodies to be punished, nor souls to be condemned, they therefore do as they like." The concept of person in many cultural, philosophical, religious, and spiritual traditions of the world including the Judeo-Christian traditions on which most of modern conceptions and imaginations of law and society exist has not only a functional dimension of performance of role but also a transcendental dimension. But the tragedy of modern law and society is that, while corporations are granted legal rights of persons, the transcendental aspect of personhood is rarely explored. But the discourse of corporations is used by owners to appropriate wealth and resources creating havoc, inequality, and suffering in ecological, social, cultural, and spiritual environments. With this foundational limitation, there has emerged the discourse of CSR. In India, there have been laws about CSR where corporations above a certain ceiling of profits are required to invest it in socially beneficial causes or donate it to actors and movements working on it. The Company Act, 2013 of India requires companies of having a net worth of more than 500 crores, turnover of more than 1,000 crores, and net profit of more than 5 crores annually are required to spend 2% of their profits in socially useful welfare and community development projects such as in health and education. This is legally required and companies are required to file their Corporate Social Responsibility (CSR) compliance report.

CSR reflects a discursive shift in the conception of corporation such as Milton Friedman's that the social responsibility of business lies in making profit for business.[1] But this is questioned by Oliver Hart and others who talk about stakeholders' welfare and well-being rather than just profit maximization for their shareholders. For example, Oliver Hart and Luigi Zingales in their work talk about maximizing shareholder welfare rather than just maximizing profit (Hart and Zingales 2017). They take issue with Milton Friedman, who says:

> That responsibility is to conduct the business in accordance with their desires, which will generally be to make as much money as possible while conforming to the basic rules of the society, both those embodied in law and those embodied in ethical custom.

Here Hart and Zingales urge us to realize:

> We argue that it is too narrow to identify shareholder welfare with market value. The ultimate shareholders of a company (in the case of institutional investors, those who invest in the institutions) are ordinary people who in their daily lives are concerned about money, but not just about money. They have ethical and social concerns. In principle, these could be part of the "ethical custom" Friedman refers to, but does not elaborate on. Not only do shareholders give to charity, something Friedman

discusses at length, but they also internalize externalities to some extent. For example, someone might buy an electric car rather than a gas guzzler because he or she is concerned about pollution or global warming; she might use less water in her house or garden than is privately optimal because water is a scarce good; she might buy fair trade coffee even though it is more expensive and no better than regular coffee; she might buy chicken from a free range farm rather than from a factory farm; etc. As another example, many owners of privately-held firms appear to care about the welfare of their workers beyond what profit maximization would require. However, if consumers and owners of private companies take social factors into account and internalize externalities in their own behavior, why would they not want the public companies they invest in to do the same? To put it another way, if a consumer is willing to spend $100 to reduce pollution by $120, why would that consumer not want a company he or she holds shares in to do this too?

(Hart and Zingales 2017: 248)

In this context, Piet Strydom tells us how, since the 1990s, there has arisen a discourse of CSR. Strydom (forthcoming) writes:

Since the 1990s, the corporate sector has been making gestures towards embracing responsibility in environmental and social respects, while in the course of time, depending on circumstances and related criticisms directed at it, ups and down in its commitment can be observed. Over the years, business network titles such as "Responsibility Inc.", "Business for Social Responsibility", "Responsible Care", "Investor Responsibility", "Corporate Social Responsibility" and more recently "Corporate Social Responsibility Newswire" and "Global Reporting Initiative" popped up, disappeared or became substituted by some new version. The themes of the annual meetings of the World Economic Forum, whose membership includes all the major corporations, reflect this corporate concern. In 1999, the Forum addressed the theme of "Responsible Globality", in 2005 "Taking Responsibility for Tough Choices", in 2017 "Responsive and Responsible Leadership" and at its 2020 meeting "Stakeholders for a Cohesive and Sustainable World". The aim is ostensibly to portray a step towards a new type of responsible capitalism, most recently for example shifting from an emphasis on shareholders to stakeholders and sustainability. On close inspection, the sincerity of some in the corporate world cannot be doubted, yet observers of the sector consistently find, in the phraseology of one of them, that "their words are bigger than their actions". Third party investigations show further that the widespread embrace of the title "corporate social responsibility" is not just confusing and therefore misleading, but in fact forms part of a broader strategy of obfuscation. Years ago already, Greenpeace exposed corporate efforts towards environmental responsibility as "greenwash". As regards the internally undifferentiated blanket expression "corporate social responsibility", while signalling an outward orientation towards others and the wider world, for many if not most in the corporate sector it in fact means a focus on the corporation itself and its contribution to society strictly as a corporation engaged in business. It is reported that, according to a spokesperson of a company producing polluting products, corporate social responsibility is "about the company, not its products". Many others, following Milton Friedman, simply refuse to accept the pretence of this formula, rejecting it as an entirely unnecessary yielding to a "guilt complex."

With the limitations of existing models and practices of CSR, Strydom (forthcoming) still hopes, like many of us practitioners and students of corporations and society, that this has the potential to create a new relationship between corporations and society:

> If corporations are corporate citizens, as they claim, then they can be expected to keep to their word and, accordingly, act like citizens who participate in the process of collective world-construction. . . . The animating question is whether the corporation could be transformed into a true corporate citizen forming an integral part of civil society, but a corporate citizen who is sensitive and receptive to the essential moral and ethical or spiritual meaning of responsibility.

We need to view CSR as a possible expression of a different partnership between corporations and society, and business and society.

But actual studies of practices of CSR yield us many-sided pictures. In their study of CSR conducted more than a decade ago, Gautam and Singh (2010) write:

> Very few companies have a clearly defined CSR philosophy. Most implement their CSR in an adhoc manner, unconnected with their business process. Most companies spread their CSR funds thinly across many activities, thus somewhere losing the purpose of undertaking that activity. Special CSR initiatives were taken by some companies like structured CSR etc. Generally speaking, most companies seem either unaware or don't monitor their company's CSR. However, all companies can be considered to be an upward learning curve with respect to CSR. The overall approach still seems to be driven by philanthropy rather than integrating it with business as has been happening in the west.
>
> (Gautam and Singh 2010)

CSR is also being put into practice in international trade. In his study of implementation of CSR guidelines in export-oriented garment industries in Tirupur, Tamil Nadu, Geert de Neeve (2009) tells us that this hides inequality between the more powerful compliers who are the buyers and the producing companies on the ground:

> While the image of a partnership was repeated to me by several buyers, their descriptions of how inspections are carried out in the factories reveal a different picture: one of stark inequality, in which buyers' moral superiority and social responsibility is contrasted with suppliers' lack of understanding, social consciousness and ethical concern. The buyers take on the role of teacher, the supplier is depicted as the apprentice, who has to be taught, disciplined and tested.
>
> (de Neeve 2009: 68)

de Neeve further tells us:

> Suppliers are not only presented as highly unreliable and untrustworthy, but also as lacking technical know-how and unwilling to learn. Buyers consider themselves superior on at least two fronts: their technical ability and knowledge on the one hand, and their moral convictions and social responsibility on the other hand.
>
> (ibid: 69)

The Companies Act of India, 2013 and Corporate Social Responsibility

CSR in India in recent times is governed by clause 135 of the Companies Act, 2013, which was passed by both Houses of the Parliament, and had received the assent of the President of India on August 29, 2013. The CSR provisions within the Act is applicable to companies with an annual turnover of 1,000 crores INR and more, or a net worth of 500 crores INR and more, or a net profit of five crores INR and more. Companies may implement these activities taking into account the local conditions after seeking board approval.

The Companies Act of India, 2013 makes India the only country in the world where CSR is mandatory. The Companies have to file their report in a specified format detailing activity-wise reasons for spending under 2% of the average net profits of the previous three years and a responsibility statement that the CSR policy – implementation and monitoring process – is in compliance with the CSR objectives, in letter and in spirit.

CSR is now linked to sustainability, and in the revised guidelines, the thrust of CSR and Sustainability is clearly on capacity building of communities and corporations for sustainability. The revised guidelines emphasize on CSR work for development of backward regions and upliftment of the marginalized and under-privileged sections of the society. This has the potential for contributing significantly in the long run to socio-economic growth in all the backward regions of the country.

In their study, "Corporate Social Responsibility Practices in the Time of COVID-19: A Study of India's BFSI Sector," Archana Koli and Rutvi Mehta

(2020) tell that reporting rate among eligible companies is around 64%. They further tell us, "Listed companies in India spent Rs. 10,000 crores in various programmes—ranging from educational programmes, skill development, social welfare, healthcare and environment conservation." They also tell us:

> The Prime Minister's Relief Fund saw an increase of 139 per cent in CSR contribution over last one year (2019–20). The education sector received the maximum funding (38 per cent of the total), followed by hunger, poverty and healthcare (25 per cent), environmental sustainability (12 per cent) and rural development (11 per cent).

On the wake of COVID-19, the Government of India is encouraging companies to spend on COVID-19-related works such as fighting hunger and malnutrition, providing health care, and creating health infrastructure.
Koli and Mehta tell us:

> Most companies use CSR practices as a marketing tool and many are only making token efforts towards CSR in tangential ways such as donation to charitable trusts, NGOs, sponsorship of events etc. Few have a clearly defined CSR philosophy. Mostly, companies implement CSR in an ad-hoc manner, unconnected with their business process and don't state how much they spend on CSR activities.

In their study, "Why This Is Not CSR: A Study of Five Major Corporates," Sristi and Tavleen Singh (2020) tell us:

> Major choices on which project to fund are made often around short-term projects which are sub-optimal to solve the social issue. In the wake of the novel coronavirus pandemic, a large proportion of the amount prescribed to CSR activities has been allocated towards the PM-CARES Fund, specifically, Rs 5,324 crore. This despite the fund being rife with issues pertaining to transparency and accountability, along with a general lack of clarity over the control and use of money donated to it.
>
> Moreover, companies have traditionally preferred to build physical structures like hospitals and schools because apart from being, quite literally concrete, they can also carry branding. Though these structures are necessary, by doing this, they distance themselves from the field and hence, the activities tend to be a form of charity rather than following major principles of grounded social work.

In their study, they discuss partnership between Infosys Foundation and Akshaya Patra Foundation, an initiative of ISKON which provides lunch to schools in Karnataka. In this context, the authors tell us:

> In Karnataka it was seen that the meals provided by them were not being consumed by the children as they were not made of staple ingredients

and diet preferences of the community. Instead it followed a *"satvic"* diet, with no garlic, onion or eggs, following the ideas of ISKON.

This is a perpetuation of the notions of purity associated with the "upper castes". If the children are considered actual stakeholders, their needs, preferences and culture would be a priority and this might have been picked up, however with a decision-making committee composed mostly of Brahmin men, and a top-down approach to welfare, this probably did not even register as a concern.

In their study, they also study Reliance Foundation's work on health care. They write:

> The foundation has supported multiple cornea transplants under the "Drishti" project. All these interventions make clear the techno centric and tertiary level focus of the foundation. If efforts are made to bring health care to the rural areas, it is mostly to connect them to the services of the city.
>
> This is an unsustainable and narrow-minded approach to health, which does not take into account the WHO definition that clearly states that health is not merely the absence of disease. It has multiple social determinants which frame it including an individual's social location (sex, class, caste being some of them), and access to basic services such as housing, clean water and food.
>
> The foundation has vertical programmes that invest huge sums of money in tackling specific diseases but fails to see the interrelations of various social issues that act as upstream factors that actually lead to the disease.
>
> It tackles tuberculosis while not looking at issues of poor ventilation and housing that cause the disease to flourish. It provides counselling to rural women about anemia without taking into account the biases in society that cause her to be the last person to eat a meal in her household, often incomplete and lacking all nutrients. The foundation acknowledges the rise in diabetes but does not link it to the growing unregulated processed food market that has infiltrated all areas and strata of society, increasing profits for food companies but causing these diseases in the population.
>
> These vertical programs do not look at health as a comprehensive issue that also has preventive and promotive components. It does not plan ways to train the health care community workers themselves and empower the community itself.
>
> It looks at beneficiaries as passive patients who must accept the treatment provided, mostly by private interventions, to them. Health is then not a right but a charitable gift. Its initiatives do not promote ideas of self-reliance, behavioral changes or a sense of ownership by the community.

Some other critical studies on CSR tell us about the criminal collusion between State and corporations in supressing worker's and people's rights and unleashing terror and violence which is not just an instance of corporate irresponsibility but corporate terror. Here the work of Sterlite Industries in Thoothukudi, Tamil Nadu, is a case in point. On May 22, 2018, police opened fire on people protesting against the functioning of Sterlite in their areas killing 15 people. Many others were also injured by the lathicharge of the police. It is a case of "The Looming Corporate Will" which leads to suppression of democratic rights of the people and eventually suppression and killing of people. As K. Rajavelu and Stanley Joseph write:

> Ever since its initiation in 1995, the Sterlite plant was an issue of contention at it continued to allude norms and compliances, which impacted the lives and health of people. The decision to expand in the face of such opposition had led to popular unrest in the community. They were further frustrated because despite local and national judicial and administrative bodies having documented water contamination, air pollution and other forms of environmental degradation linked to the copper smelting plant and related activities, the state had not taken cognizance of the matter and seemed to have colluded with the Vedanta group.
>
> It is pertinent to note that it was the Company that first claimed the perceived threat based on which the action of the state followed. The Company, instead of seeking protection for itself, if it really apprehended danger, instead sought the banning of the democratic right of the people to protest in a public space. The police complied and this led to severe consequences for the people. All this was happening while the people of Toothukudi had clearly communicated that their protest was not against all industries, but against the hazardous ones with unscrupulous practices, affecting the people and the environment. The government of Tamil Nadu had also failed to point out before the Honourable Madras High Court the huge impropriety of a private company seeking a ban on any democratic activity in the public space to safeguard its private interests. This infringement on the rights of the community to protest peacefully, seeking answers and redressal of issues of non-compliance by the Company, was a direct attack on the fundamental rights as enshrined in the Constitution of India.
>
> Even a year after the mass reprisal and ongoing human rights violations by the state at the behest of the corporate, the Company continues to mark its presence in the region through its CSR activities. And all this is happening while the local administration seems to be playing the role of "watch-dog" on behalf of the Company.

The situation in Thoothukudi very well underlines a dangerous and growing reality in India where companies leverage the government and its agencies to use force to subvert popular voices against them and their unethical operations as an extension of corporate will.

(Rajavelu and Joseph 2019: 71)

Taking into consideration violence in Thoothukudi and the septic role of CSR there and other places in place of entrenched corporate irresponsibility, Amita Joseph et al. (2018) write:

> Corporate irresponsibility can also, in varying ways, cost lives. The murder of 13 protestors, including a 17-year old student in Thoothukudi, Tamil Nadu, in May 2018, was a brutal example of State aggression towards an unarmed crowd. The crowd was trying to build awareness of the wider society on environmental pollution and consequent health hazards to local villagers caused by M/s Sterlite.
>
> Today, business leaders are garlanded at "CSR" awards as if they are doing something much more than complying with the law; but no 2 percent CSR story, however positive, should be allowed to camouflage corporate irresponsibility. India's corporations have a real opportunity to redress social inequalities, to transform the conditions of their workers and those working across their value chains, to safeguard the human rights of local communities as well as to ensure that the environment is protected: indeed, of contributing to national development, as envisaged in the National Voluntary Guidelines. But are they ready for responsibilities that this entails?

In the previous paragraph, Amita Joseph et al. tell us how corporations really do not take steps towards redressing problems in society such as social inequality, poverty, and collusion of state and corporations in creating and perpetuating violence in society.

Corporate Spiritual Responsibility

The discussion so far presents us some glimpses of discourses and practices of CSR in India. CSR follows most of the time a trodden path of investment or philanthropic donation in social sectors such as health, education, and infrastructure. Even there, concerned actors are not interested in the outcome of their investments. There is need for much more improvement and transformation on this front as well. At the same time, corporations as multi-dimensional agents in society can also contribute to addressing the aesthetic and spiritual challenges of society. Indian society is a multi-cultural and multi-religious society and conflicts often break out around religious issues. Such conflicts and violence as during Gujarat riots of 2002 and anti-Sikh riots in Delhi in 1984 affect the social, economic, moral, and spiritual well-being of communities, polity, and economy of our nation. But corporations do not feel the need to work in the field of inter-religious dialogue and communal harmony. To address such challenges is an aspect of spiritual responsibility of corporations. Another related work is aesthetic responsibility of corporations where corporations can try to create aesthetically rich environment which would inspire people to think differently. An example of such aesthetics of responsibility is the existence of public sculptures and art in a city like Munster which brings symbols from different religions of the world together. Aesthetic corporate responsibility can bring new dimensions to drive for cleanliness in contemporary India along with governmental campaigns such as Swachh Bharat. This can also be part of a broader agenda of rethinking and transforming development as aesthetic and spiritual and not just economic and political (see Clammer and Giri 2017).

Against this backdrop, we need to explore further the vision and practice of corporate spiritual responsibility which can help us transform existing conceptions and practices of corporations, responsibility, CSR, state, market, self, and society. Here we need to explore frames and ways of responsibility which go beyond existing polarities such as state and market, and market and NGOs and brings us to fields, circles, and spirals of responsibility as a process in which all these constituent actors co-present and interact with each other in a spirit of autonomy, interaction, interpenetration, and transformation (Quarles van Ufford and Giri 2003). The discourse and practice of corporate spiritual responsibility can transform the existing discourse of CSR by cultivating a spiritual dimension within and across self, corporations, organizations, and society. It can help us explore cases of corporations and service delivery groups which follow a spiritual approach to self, society, and market by pointing to the need for cultivating the dimension of love, care, and responsibility—our potential real higher self—and not just be a slave of narrow self-interest. We find such spiritual approaches in many contemporary visions and practices such as Muhammad Yunus' (2008, 2010) vision of social business, C.K. Prahalad's (2004) approach of partnership with the poor beyond the demeaning and paternalistic language of poverty, Subhash Sharma's (2012) approach of rotating the pyramid with capillary actions, and the work of Arvind Eye Hospitals initially working in Madurai and now in other parts of India for creation of joy, prosperity, and beauty for all.

As is well known, Muhammad Yunus is the nurturer of Grameen Banks in Bangladesh which created a space for dignity and development for the resource poor women and men of Bangladesh. From his experience and experiments, he launched the idea of social business where entrepreneurs do business on themes and issues which are of need to society and which are usually neglected by the mainstream dominant economic interest (Yunus 2010). In social business, concerned actors and entrepreneurs do business on issues which fulfils an unmet need of a society which in turn also gives an opportunity to them to earn a livelihood and achieve many-sided growth and development. Social business helps us to realize that entrepreneurs have a reality of concern for others in them and are not solely interested in profit maximization. Thus, social business in Yunus is ultimately based upon a spiritual view and realization that self, society, and economy have a concern for others as well as concern for self. This goes beyond a naïve dualism between altruism and egoism and is simultaneously interested in well-being, growth, and development of self and others. Similarly, Prahlad's engagement with creating business services and deliveries for people at the bottom of the pyramid also has an implicit and yet-to-be-realized spiritual dimension. It is born out of the fact that we have to go beyond paternalism and create business services for the people who are at the bottom of the pyramid. Subhash Sharma (2012) here argues that it is not enough to look at people as existing at the bottom of the pyramid, we need to conceptualize and realize them as agents of dignity and we also need to rethink and transform the very pyramidal structure. We need to rotate the pyramid itself. Pyramids represent many a time hierarchy and top-down control. They can be pyramids of sacrifice where well-being and creativity of people may be sacrificed. It is in this spirit Sharma talks about the need to rotate the pyramid (also see Sharma 2007). Sharma here draws our attention to the visions and works of capillary actions in societies and histories which change the existing structures of pyramid. We find such capillary actions in the works of NGOs, social movements, and other change agents in market, corporations, state, and civil society. But along with capillary action which refers to grassroots movements for participation and creativity, we can also add capillary meditation as an important part of rotating the pyramid and creating movements of responsibility. Corporate spiritual responsibility challenges us to rethink and transform human action including in fields such as market and business as simultaneously active and meditative, animated by meditative verbs of co-realizations which bring action and meditation together (Giri 2012).

In his works, Subhash Sharma adds further dimensions to the discourse and practice of corporate spiritual responsibility. In a recent article on this, Sharma (2020: 13) writes:

> The idea of Corporate Social Responsibility (CSR) is now well accepted by the corporate world. Further the idea of spirituality in management has also been finding some acceptance among some corporate leaders. In fact social responsibility at a deeper level is an extension of spiritual responsibility. For defining spiritual responsibility we should have a workable definition of spirituality. This leads to the idea of religion (s), connectivity with nature leads to the idea of science, connectivity with Self leads to the idea of spirituality and connectivity with cosmos/cosmic consciousness leads to the idea of enlightenment.

Sharma here presents a holistic framework for corporations where these are simultaneously concerned with ecology, equity, ethics, and efficiency. Sharma brings consciousness to work with ecology, compassion with equity, cooperation with ethics and competition to efficiency. Corporate spiritual responsibility involves bringing consciousness, compassion, cooperation, and competition to work with ecology, equity, ethics, and efficiency.

Corporate spiritual responsibility strives to rethink corporations and individuals as part of an ecology of relationships inhabited by human, social,

Nature, and the Divine—all understood in a very broad sense and not in a narrow way going beyond any particular dogmatism and absolutism such as anthropocentrism, natural determinism, and authoritarianism in the name of God. It calls for realizing the multi-dimensional realities of immanence and transcendence in our lives. Corporate Spiritual Responsibility calls for development of a new vision and practice of leadership which touches the soul/ Atmic dimension of self, corporation, and society as well as the bio-regional and planetary dimensions. Leadership here is Atmic-Soul nurturing and Planetary. It builds upon the contemporary ideas of bio-regional leadership and strives to transform this in the direction of Atmic Planetary Leadership where leaders of corporations try to attune themselves to their souls, listen to its calls for Infinity with humility and courage, and do not get trapped in narrow battles of ego, power, and profit (see Howard 2019). They also become bio-regional by listening to the rhythms of their bodies, soul, and soil and developing programmes of development which build upon the bio-regions of the locality and community rather than just exploit the resources of local communities for profit maximization of state, market, and corporations leading to the degradation, denudation, and annihilation of self and society in local communities (Howard 2019). In this, they pay attention to preserving and nurturing bio-diversity of bio-regions as well as the world. They prepare corporations to invest and do business in such ways which contribute to preservation and flourishing of bio-diversity rather than destroy it.

Corporate spiritual responsibility challenges corporations to focus on the spiritual challenges of their own employees, society, and humanity. Many of the employees of corporations are going through pain, profound spiritual poverty, and crisis of meaning. Here what Sharma (2020: 18) writes deserves our careful consideration:

> Ajeet Saxena . . . suggests the need for developing a metric for the spiritual health of corporations. He suggests that only spiritually awakened person should be given assigned to responsible positions in the organization. "Spiritually challenged" individual should necessarily start at the lower level. Every employee should undertake "Spiritual MRI" by way of Aura mapping and his/her "Spiritual Lipid" profile should be measured by measuring composition of tamas, rajas and sattva qualities.
>
> Spiritual health of an organization can be measured by assessing the intensity of negergy (negative energy) and synergy in organization culture. If negergy is high spiritual health is low and there is a need for a "spiritual alignment" within the organization. This can be done by undertaking Human Quality Development (HQD) programmes for the employees and introducing the concept of Spirituality in Management.

Existing counselling and therapy programmes are inadequate to address the problem of spiritual illness of self, corporations, and society now. Corporations need to put their own houses in order by transforming them into ecologies and gardens of well-being and creativity and not just money-making machines and blood-thirsty wheels. Corporations work with the same logic, teleology, and theology of power where to rule and control runs supreme. It is within the framework of rules and kingdom and it can cultivate a different political theology and practical spirituality of gardening with Gods rather than being under or with the regnant discourses and theologies of Kingdom of God (Giri 2021). Corporations then can take up such challenges in wider societies such as need for creating initiatives for peace and harmony across cultural, religious, and ethnic divides which are usually not taken up in existing programmes CSR. Corporations can also help us to go beyond a preoccupation with money and profit maximization and help us realize Artha—both money and meaning—as part of a wider and deeper search and realization of meaning in self, society, and the world (Giri 2019a).

Climate Change, Anthropocene, and the Calling of Responsibility and Socio-Spiritual Transformations

> Responsibility has a double face: on the one hand, we need to be very clear about which human activities continue to contribute most to climate change while, on the other hand, not ignoring *the generic or diffuse sense* in which everyone is implicated, as everyone must be. In the first case the channels along which "responsibility" runs are invariably pre-figured, already cut into shape, and climate activists among others have to focus on re-shaping them. [. . .] The second raises more general questions about how people lead their lives, and the very contours of the world within which taking responsibility makes any sense; there are also questions about what precisely is understood to be "human" when commentators point to the enduring impact of the Anthropocene, that is, human intervention evident as a geological stratum,. We are. in effect, talking cosmologies.
>
> Marilyn Strathern (2021), "Regeneration and
> its Hazards," pp. 1–2

Climate change is a major challenge of our times. With climate change, there is global warming which is leading to rise of temperature and sea level with inundation of landmass, habitats, and consequent despair and displacement (see Eriksen and Schober 2017). Climate change is a product of over extraction and unsustainable use of natural resources leading to greater degree of CO_2 emission. The Intergenerational Panel of Climate Change (IPCC) had prepared scientific evidence about climate change and global warming in the late 1980s. The Earth Summit in Rio de Janeiro in 1992 tried to address some of these issues and created the following documents: Convention on Biological Diversity, Framework Convention on Climate Change, and United Nations Convention to Combat Desertification. After this, Kyoto Protocol was signed in Kyoto in 1997 to create a legally binding framework for addressing the problems of climate change. The Kyoto protocol was an international treaty which extended the 1992 United Nations Framework Convention on Climate Chante (UNFCC) that commits state parties to reduce greenhouse gas emissions. This protocol entered into force on February 16, 2005. This was followed by the Copenhagen Accord of 2009 which was then followed by Paris Agreement of 2015. The Paris Agreement is a legally binding international treaty on climate change. It was adopted by 196 Parties at COP 21 in Paris, on December 12, 2015 and entered into force on November 4, 2016. The Paris Agreement is a legal agreement which binds all singing nations to take concrete steps and commit themselves to reduce CO2 emission and other climate change related actions.

Paris Agreement uses market-based approaches to address the challenge of climate change such as Payment for Environmental Services (PES) and Markets for Environmental Services (MES). As Maria Natalia P. Rodriguez writes, "economic, social and environmental criteria . . . were only marginally addressed and were not integrated in the objective or operational guidance set forth in the Paris Agreement on PES and MES" (2019). Other commentators such as British climate scientist Dr. Kevin Anderson challenges us to understand the "implicit savagery underlying the Paris accord, as he envisioned rich people in the global North muddling through an coping with the climate crisis as it unfolds, while the poor people in the global South were expected to die off" (Rao 2016b: 25). The Paris Agreement also seeks to address the question of carbon emission through carbon trade, but critical scholars such as Jamie Morgan and Heikki Patomaki here propose for carbon tax or greenhouse tax not only through nation-states but also through some supra-national, transnational bodies (see Morgan and Patomaki 2021). Morgan and Patomaki write:

> Atmospheric greenhouse gases (GHGs) or more generically "carbon emissions" do not respect borders and so "decarbonisation" is everyone's problem. It is either achieved everywhere or undermined globally. It is, in a literal sense, a planetary problem. Not all places and people are equal sources of, or are equally responsible for, carbon emissions, and not all people and places are or will be able to equally buy their way out of some of the immediate consequences, but in the end climate change is a pervasive existential civilizational challenge. This creates scope for global responses, including a global greenhouse gas tax (GGGT).
>
> (2021: 1)

For Morgan and Patomaki,

> carbon taxes should be preferred to carbon trading, and second that there is a need for the organisation and coordination of this taxation at a global level. Moreover, there is also a case to be made for a market-disruptive rather than market-conforming approach to carbon taxes.
>
> (ibid: 3–4)

The Paris Accord had some ups and downs. Donald J. Trump upon his election as President of the USA in 2016 pulled the USA out of the Paris

Agreement on Climate Change. President Joe Biden upon his election as President of the USA in 2021 brought the USA back to Paris agreement with a renewed sense of commitment. Other countries such as China and India also seem to show concern for this. But Paris Accord is not our panacea. It is still too top down and our current climate catastrophe calls for many-sided actions from all concerned. Here creative people and organizations across the world are coming together to create green covers and forests. They are part of what M.S. Swaminathan (2011) calls climate care movement. Ashish Kothari, Arturo Escobar, and friends associated with projects such as pluriverse and Vikalp Sangham Tapestry of Alternative Processes present us such a view (see Kothari et al. 2019).

The latest IPCC report on Climate Change released on August 9, 2021 stated that the Earth has been hottest now in the last 125,000 years. This calls for responsibility—global responsibility. In 2018, Greta Thunberg, a young Swedish girl, came and started sitting in front of the Swedish Parliament in Stockholm on Fridays leaving her school to draw the attention of the lawmakers of the country to address the problem of climate change. Thunberg's protests for three weeks had an effect. This then gave rise to the global movement of Fridays for the Future in which school children across the globe took part before the onset of COVID-19. In her many speeches, Greta Thunberg tells us that our civilization is so fragile. "It is a castle built on sand." The elected leaders and decision-makers should hear the coming of imminent collapse and take emergency steps to avert climate disaster. Greta tells us that "we are in the beginning of a mass extinction." For her, "the climate and ecological crises is beyond party politics and we cannot make deal with physics." Addressing the leaders and decision-makers, she says, "You must do the impossible because giving up can never be an option." Greta and students involved in Friday for the Future challenge the present generation to realize: "If you choose to fail us, we will not forgive you." For Greta Thunberg, "No one is too small to make a difference."

Greta Thunberg is painfully aware that people think that she is "retarded, a bitch and a terrorist." But she is not bothered by this and she wants us to think beyond the box what she calls "cathedral thinking." In one of her speeches, Greta tells us,

> Adults keep saying: we owe it to the young people to give them hope. But I don't want your hope; I don't want you to be hopeful. I want you to panic, I want you to feel the fear I feel every day. And then I want you to act; I want you to act as if you would in a crisis. I want you to act as if the house was on fire, because it is.

In her article, "Climate Change: A Ray of Hope," Bindu Mohanty (2019) writes about Greta Thunberg and *Friday for the Future*:

> The #FridaysforFuture movement has gathered rapid momentum since it started. After Thunberg addressed the 2018 United Nations Climate

Change Conference in Poland, student strikes took place every week somewhere in the world. Then, in 2019, during the worldwide Global Climate Strike from September 20–27 millions more, students and adults, joined the movement. In 2019, there were at coordinated multi-city protests involving millions of people, and on September 23, Thunberg galvanized the world when she addressed the UN Climate Action Summit with a short, emotional speech stating:

This is all wrong. I shouldn't be up here. I should be back in school on the other side of the ocean. Yet you all come to us young people for hope. How dare you! You have stolen my dreams and my childhood with your empty words. And yet I'm one of the lucky ones. People are suffering. People are dying. Entire ecosystems are collapsing. We are in the beginning of a mass extinction, and all you can talk about is money and fairy tales of eternal economic growth. How dare you! For more than 30 years, the science has been crystal clear. How dare you continue to look away and come here saying that you're doing enough, when the politics and solutions needed are still nowhere in sight.

Mohanty relates Fridays for the Future movement to other current movements such as the "Occupy Movement" and # Me Too Campaign. As Mohanty writes:

It was left to Greta Thunberg, however, to author such a massive global movement through the #FridaysforFuture campaign. The Thunberg effect of widespread solidarity for a cause is similar to other peoples' movements such as the "Occupy Movement" and social media campaigns, notably the #MeToo campaign, which led to cultural changes in the world and threatened the status quo. What is singular about the #FridaysforFuture movement and climate change activism, however, is that it seeks to tackle an issue, which affects not just one particular oppressed sector of society but the entirety of humanity. Never before has such a cause as climate action, to curb our carbon emissions and limit global warming to less than 1.5 degrees Celsius as recommended by IPCC, united diverse human populations all over.

Extinction Rebellions is another contemporary movement against climate change which uses non-violent ways such as stopping traffic in busy streets to draw the attention of the public and concerned authorities to take action for climate change. It urges us to realize that we are face to face with the Sixth Extinction in Earth's history where human species would be extinct. The initiative was taken by molecular physicist Gail Bradbrook and social science professor Roger Hallam. The Mobilising Song for *Extinction Rebellion* is:

People gonna rise like water
Gonna turn this system round
In the name of my grand daughter.

Like Fridays for the Future, Extinction Rebellion urges us to realize and embody our responsibility across generations—generational responsibility towards preserving and nurturing our Mother Earth—and to save humanity from its imminent climate disaster and collapse.

Anthropocene

Climate change is happening in the changed geological stage of our planet called Anthropocene. Anthropocene marks an epoch in geological history of humanity which is influenced by human foot prints. It has given rise to aforementioned problems of climate change.

The discourse of Anthropocene began from scientific fields such as geology and is now spreading from a narrow scientific discourse to a wider cultural concern and reflection. House of World Culture, Berlin, had organized an international conference on the challenge of the Anthropocene in Berlin in January 2013. It brought scientists, artists, philosophers, scholars of literature, and interested common people together. There were also novel experiments like creative kitchen and preparing food with sensitivity to the challenge of the climate change. There were also novel forms of mutual exploration such as organizing sessions where the leader brought an object and reflected upon the challenge of climate change and Anthropocene through that object. One speaker had brought a cat to reflect upon the challenge of the Anthropocene by pointing out that with the Anthropocene there is a distinct possibility that human beings would be wiped out and only species like cats would remain. Another speaker, an artist, came on stage, bared himself, and started crawling like an animal to show the way human beings would be extinct with the coming Anthropocene.[2] After the Berlin conference on Anthropocene, there was also a seminar at Munich on the issue of cultural dimensions of responses to Anthropocene in June 2013 in which I had taken part. It was organized by Rachel Carson Center of Ludwig Maximilian University (LMU), Munich.

Dipesh Chakraborty was one of the participants in the 2013 House of World Culture Conferences on the Anthropocene. I had met with him there. Before this Chakraborty had written a widely acclaimed essay in 2009 called "The Climate of History: Four Theses." Now Chakraborty has built on this reflection and written *The Climate of History in a Planetary Age* (2021). For Chakraborty, the present moment of Anthropocene challenges us to understand the distinction between the globe and Planet Earth. For Chakraborty, "the globe is humancentric . . .; the planet, or the Earth System, decenters the human" (2021: 4). The condition of Anthropocene presents a condition of "shared co-responsibility" of humans and non-humans. For Chakraborty,

> as humans we presently live in two different kinds of "now-time" simultaneously in our awareness of ourselves, the "now" of human history

has become entangled with the long "now" of geological and biological time scales, something that has never happened before the history of humanity.

(Chakraborty 2021: 7)

For Chakraborty, the globe in globalization is different from the globe of global warming. The globe of globalization is produced by histories of colonial violence and neo-liberal capitalism while the globe in global warming is related to our current planetary condition and our responsible thought and action to continue to keep our planet habitable. In the discussion title in his book, "Sustainability and Habitability: Distinguishing the Global From the Planetary," Chakraborty tells us that sustainability belongs firmly to "the history of the global" (p. 81) while "the anthropocentric idea of sustainability dominated the twentieth century" (p. 82). For Chakraborty, "The geological time of the Anthropocene and the time of our everyday lives in the shadow of global capital are intertwined" (ibid: 10). While global capitalism as part of global industrialism which includes earlier periods of global socialism has contributed to the making of the Anthropocene, the challenges before us is how to think and act which "helps humans to be at home on earth beyond the time of living" (ibid: 9). For Chakraborty,

> A consumption-driven capitalism in which all artifacts are up for consumption in the present would be an antipolitical machine in that it would eventually work against the logic of human dwelling, since dwelling requires artifacts beyond the lifetime of living.
>
> (ibid: 9–10)

The Anthropocene moment, for Chakraborty, urges us to realize the "agency of the non-human and the non-living" (ibid: 100). In the age of the Anthropocene, we have a responsibility to belong to the planetary in the spirit of the planetary where we are all inhabitants of our planet which calls for us to go beyond our conventional and existing anthropocentric domination. Here Chakraborty urges us to realize that "it is not always possible for humans to transition smoothly from being attached to a human-dominant order of life to being one species among many" (ibid: 113). Political, cultural, and spiritual movements can help us in this transition embodying what Martha Nussbaum (2006) calls "cross-species dignity" (also see Gilbert 2018). In his work, Chakraborty refers to the works of Pope Francis and Amartya Sen who build on Christianity and Buddhism to embody our responsibility to Nature and living responsibly so that we do not harm human species as well as other species.[3]

In our initial discussion of global responsibility, we had spoken about aesthetic dimensions and ways of global responsibility, and similarly, we can cultivate aesthetics of the Anthropocene which can help us realize the link between human and non-human as part of Nature. Here what Eva Horn and Hannes Bergthaller (2020) write deserves our careful consideration:

> An aesthetics of the Anthropocene, we believe, need to deal not so much with the alienation of humans from nature but with a more thorough-going alienation—the becoming uncanny of the lifeworld . . . in the Anthropocene nature can no longer be understood as a collection of inert objects, as a world of passive matter which is perceived and acted upon by an observing (human) subject but itself incapable of perceiving and acting, a world that occasions reflection but reflexivity. On the contrary, as Amitav Ghosh has argued, nature now returns the gaze—it appears alive, threatening, unpredictable, sentient and temperamental. As Ghosh writes, this is "one of the uncanniest effects of the Anthropocene, these renewed elements of agency and consciousness that humans share with many other beings, and even perhaps with the planet itself."
>
> (Horn and Bergthaller 2020: 101)

For Horn and Bergthaller, this moment calls for an affirmative biopolitics as part of new realization of cross-species dignity as they write:

> Neither the biopolitics of scarcity nor the biopolitics of abundance seems to be able to point towards a livable future, then. Ultimately, this leads to the question of whether a different form of biopolitics can be imagined—a biopolitics which would not oppress or control life but contribute to its genuine flourishing.
>
> (Horn and Bergthaller 2020: 124)

For Horn and Bergthaller,

> What is at stake in the Antropocene is above all the *global* agency of the human as *anthropos*. The space in which the human is situated now is no longer that of a local environment, but that of the world as a whole.
>
> (Horn and Bergthaller 2020: 146)

Being part of the world as a whole calls for us to cultivate planetary mode of being with the world which needs to interact with the global mode and then transform it. Chakraborty tells us about two ways of being with our world—the global and the planetary. But these are two interlinked modes with realities and possibilities of mutual critique and transformations. As Chakraborty writes:

> For all their differences, thinking globally and thinking in a planetary mode are not either/or questions for humans. The planetary now bears

down on our everyday consciousness precisely because the acceleration of the global in the last seventy years or so years—all that is summed in the expression "the great acceleration"—has opened up for humanist intellectuals the domain of the planetary.

(ibid: 85)

Chakraborty builds upon philosopher Karl Jasper's idea of "epochal consciousness" to help us realize the change of consciousness and relationship that is called upon to address the challenges of climate change and the Anthropocene. It calls upon us to live ethically in our world, as he writes, "Epochal consciousness is ultimately ethical. It is about how we comport ourselves with regard to the world under contemplation in a moment of global—and now planetary—crisis. It is what sustains our horizons of action" (ibid: 197). The epochal consciousness of Anthropocene and our planetary age calls for us to be prudent, embodying prudential wisdom of mutual care, care of our Mother Earth, and responsibility. For Chakraborty, Anthropocene calls for transformation of *Homo sapiens* to be *Homo prudens* as he writes: "Can *Homo sapiens* learn to be *Homo prudens*, whatever the political battles that divide us?" This calls for prudential thinking, action, and moral courage as exemplified in the lives and thoughts of Mohandas Karamchand Gandhi and Rachel Carson who show us how to live responsibly with our Earth going beyond the dominant "visions of the human" (ibid: 202) (also see Chakraborty 2019; Mohanty 2016).

Chakraborty's work on the Anthropocene can be read together with the works of many other contemporary pioneering thinkers such as James Lovelock, Bruno Latour, Donna Haraway, Arturo Escobar, and Piet Strydom. In his work on Gaia which began with his Gaia hypothesis, Lovelock urges us to realize Earth as a living system. Lovelock urges us to see "health of the Earth as primary, for we are utterly dependent upon a healthy planet for survival" (Chakraborty 2021: 65). Here Bruno Latour urges us to realize the difference between Galilean objects and Lovelockean objects. Relating this to Chakraborty's earlier modes of the global and the planetary, we can read the following thoughts of Bruno Latour:

With Galilean objects as the model, we can indeed take nature as a "resource to exploit," but with Lovelockian agents, it is useless to nurture illusions. Lovelock's objects have agency, they are going to react—first chemically, biochemically, geologically—and it would be naive to believe that they are going to remain inert no matter how much pressure is put on them. In other words, economists may make nature a factor in production, but this would not occur to someone who has read Lovelock- or Humboldt, for that matter.

(Latour 2018: 80; also see Latour 2005)

Resonanting with Chakraborty's mode of planetary, Latour tells us that it is helpful now to re-envision and realize human beings as not just humans but as terrestrials. Latour tells us:

> It is perhaps time, in order to stress this point, to stop speaking about humans and to refer instead to terrestrials (the Earthbound), thus insisting on humus and, yes, the compost included in the etymology of the word "human."
>
> (Latour 2018: 89)

For Latour, "Terrestrial" has the advantage of not specifying the species (ibid: 86). This helps us to go beyond anthropocentric dominance and co-exist with other species with love, care, dignity, and responsibility. As terrestrials, for Latour, we need to shift from a "system of production to a system of engendering" and we can "multiply the sources of revolt against injustice and, consequently, to increase considerably the gamut of potential allies in the struggles to come for the Terrestrial" (ibid: 101). In a way, the Corona virus which emerges out of destruction of wild habitat and meeting of human species and other species becomes an ally in realizing the limitations of the existing production paradigm and engender new ways of production, consumption and living together with Mother Earth (more on this on the next section on COVID-19). Latour also tells us that as terrestrials, we co-create each other going beyond the dualism of organism and environment. For example, as Latour explains:

> If the composition of the air we breathe depends on living beings, the atmosphere is no longer simply the environment in which living beings are located and in which they evolve; it is, in part, a result of their actions. In other words, there are not organisms on one side and an environment on the other, but a coproduction by both. Agencies are redistributed. The difficulty we have understanding the role of living beings—their power to act, their agency—in the evolution of terrestrial phenomena reproduces the difficulty of understanding the phenomenon of life in earlier periods.
>
> (ibid: 76)

For Latour, the existing models and ways of the global and the local "afford us an inadequate purchase on the Terrestrial, which explains the current hopelessness" (ibid: 94). Here Latour urges us to understand the distinction between local minus and local plus as that between globalization minus and globalization plus. Local plus and globalization plus involve transformations of both the local and the global as we know and it. The local plus contains the seeds of the planetary within it. The global plus is not just the familiar logic of globalization; it embodies both the local and the planetary. Both local plus and globalization plus should also help us in multiplying our points of view. As Latour writes, "all that counts is understanding whether you are managing to register, to maintain, to cherish a maximum number of alternative ways of belonging to the world."

Chakraborty tells us about cultivating planetarity with and beyond the global and Latour tells us about multiplying our viewpoints. For Latour,

> Shifting from a local to a global viewpoint ought to mean multiplying viewpoints, registering a greater number of varieties, taking into accont a larger number of beings, cultures, phenomena, organisms, and people. Yet it seems as though what is meant by globalization today is the exact opposite of such an increase. The term is used to mean that a single vision, entirely provincial, proposed by a few individuals, representing a very small number of interests, limited to a few measuring instruments, to a few standards and protocols, has been imposed on everyone and spread everywhere. It is hardly surprising that we don't know whether to embrace globalization or, on the contrary, struggle against it. If it is a matter of multiplying viewpoints so as to complicate all "provincial" or "closed" views with new variants, it is a fight that deserves to be fought. If it is a matter of decreasing the number of alternatives regarding the existence and the course of the world, the value of goods and the definitions of the Globe, it is clear that we have to resist such simplifications with all our might.
>
> (Latour 2018: 15–16)

In the postscript to his book which contains a dialogue between Chakraborty and Latour, Chakraborty urges us to realize the need for developing planetary forms of governance in which we would have to learn how to govern taking the conditions and well-being of all species and not only human, a perspective emphasized by Donna Haraway (Chakraborty 2021: 215).[4]

Donna Haraway's works are here also important for rethinking our contemporary condition as well as raising questions to discourses such as Anthropocene. She quite creatively links movements like to alter globalization or autre-mondialization described in the second chapter of this book to the realities and responsibility of multi-species co-living. Haraway begins her book, *When Species Meet*: "There is promising autre-mondialization to be learned in relying some of the knots of ordinary multi-species living on earth" (Haraway 2006: 3). In her following influential work, *Staying with the Trouble: Making Kin in the Chthulucene*, Haraway (2016) develops further the vision and art of responsibility of multi-species living with our Earth. Haraway finds the designation of current condition of the world as Anthropocene inadequate, as too anthropocentric as she challenges us to realize: "We are humus, not Homo, not anthropos; we are compost, not posthuman" (Haraway 2016: 55). She also finds Captialocene to name the contemporary condition inadequate, as it links the current ills to the working of Capital without relating to the problems of taken-for-granted human

exceptionalism and learning how to live and die creatively with other species. Haraway offers the term Chthulucene to realize our mutually constituted life with Earth with other species. As Haraway tells us:

> Chthulucene is a simple word. It is a compound of two Greek roots
> (khthôn and kainos) that together name a kind of time place for learning
> to stay with the trouble of living and dying in response-ability on a dam-
> aged earth. Kainos means now, a time of beginnings, a time for ongoing,
> for freshness. Nothing in kainos must mean conventional pasts, pres-
> ents, or futures. There is nothing in times of beginnings that insists on
> wiping out what has come before, or, indeed, wiping out what comes
> after. Kainos can be full of inheritances, of remembering, and full of
> comings, of nurturing what might still be. I hear kainos in the sense of
> thick, ongoing presence, with hyphae infusing all sorts of temporalities
> and materialities.
>
> (Haraway 2016: 2)

Haraway urges us to realize how we have a responsibility to stay with our
troubles, as we make troubles with our current conditions of creation of pain,
suffering, unsustainability, and collapse of our world. Haraway tells us:

> Trouble is an interesting word. It derives from a thirteenth-century
> French verb meaning "to stir up," "to make cloudy," "to disturb." We—
> all of us on Terra—live in disturbing times, mixed-up times, troubling
> and turbid times. The task is to become capable, with each other in all of
> our bumptious kinds, of response. Mixed-up times are overflowing with
> both pain and joy—with vastly unjust patterns of pain and joy, with
> unnecessary killing of ongoingness but also with necessary resurgence.
> The task is to make kin in lines of inventive connection as a practice
> of learning to live and die well with each other in a thick present. Our
> task is to make trouble, to stir up potent response to devastating events,
> as well as to settle troubled waters and rebuild quiet places. In urgent
> times, many of us are tempted to address trouble in terms of making
> an imagined future safe, of stopping something from happening that
> looms in the future, of clearing away the present and the past in order
> to make futures for coming generations. Staying with the trouble does
> not require such a relationship to times called the future. In fact, staying
> with the trouble requires learning to be truly present, not as a vanishing
> pivot between awful or edenic pasts and apocalyptic or salvific futures,
> but as mortal critters entwined in myriad unfinished configurations of
> places, times, matters, meanings.
>
> (Haraway 2016: 1)

Multi-species co-living calls for what Haraway calls response-ability.
"Response-ability is about both absence and presence, killing and nurturing,

living and dying—and remembering who lives and who dies and how in the string figures of naturalcultural history" (Haraway 2016: 28). Haraway further tells us:

> We are all responsible to and for shaping conditions for multi-species flourishing in the face of terrible histories, and sometimes joyful histories too, but we are not all response-able in the same ways. The differences matter—in ecologies, economies, species, lives.
>
> (ibid: 29)

Over the last four decades, Arturo Escobar has also been engaging with the contemporary condition critically and creatively beginning with his influential work, *Encountering Development: The Making and the Remaking of the Third World* (Escobar 1995). As an anthropologist, Escobar builds upon his anthropological work on indigenous communities in Colombia as well as on traditions such as Buddhism to overcome varieties of dualisms such as Nature and Human that haunt us. Escobar (2018) calls for creating a pluriverse in place of monological worlds of various kinds that annihilate us. Pluriversal politics calls for a new relationship between Nature and Human as well as works for social and political transformations which would help us come to term with the challenges of climate change and the imminent collapse. In pluriverse, "people—relearn what it means to be a humble part of 'nature,' leaving behind narrow anthropocentric notions of progress based on economic growth" (Kothari et al. 2019: xxiii). As Haraway links multispecies living to alter-globalization movement, Escobar et al. link strivings for pluriversality to alter-globalization movement: "the alter-globalization movements propose pluriversality as a shared project based on the multiplicity of 'ways of worlding'" (Kothari et al. 2019: xxiv).

Realization of pluriversality calls for new movements in practice and thinking. Escobar (2020: 30) here heightens the need for the following movements:

> The recommunalization of social life, as a counter to the dominant individualizing imperative and as the foundation for human action from the perspective of the interdependence of everything that exists. The relocalization of activities, in the domains of economy, food, health, energy, transportation, education, building, and so forth, to resist the delocalizing tendencies of capitalist globalization, strengthen local and regional economies, and foster convivial modes of living. The strengthening of collective local autonomies and direct forms of democracy, as a means to lessen the dependence on norms established by experts and the state; critically revalorize local knowledges and values; and promote horizontal political strategies based on people's self-organization, potentially linking up with other similar transformative experiments and autonomous movements elsewhere.

For Escobar, this calls for following three broad theoretico-political movements: first, "the simultaneous depatriarchalization and decolonization of societies, as a way to move decidedly toward nonpatriarchal, nonracist, and postcapitalist social practices and organizations"; second, "the liberation of Mother Earth, as an ethical-political principle to create novel forms of existing as living beings and to rethink the relations between humans and nonhumans in mutually enhancing manners"; third, "the flourishing of the pluriverse, to weave multiple paths toward a world of many worlds, countering the power of the current model of a single globalized world and the capitalist hydra" (Escobar 2020: 30).

Pluriversal politics is ontological politics which urges us to realize the quality of our being, inter-being, interrelationship, and becoming. It is a politics of radical relationality which also has a spiritual dimension of service, care, and responsibility. Escobar's political ontology is not the political ontology of Martin Heidegger of which anthropologist Pierre Bourdieu (1991) offers a critique of its logic of mastery and silence on killing and othering of the other as it happened during Hitler's reign and Heidegger's silence about it. Escobar invites us to understand the distinction between strong relationality and weak relationality. For Escobar (2020), in weak relationality, as in dominant forms of modern politics,

> entities are first assumed to be ontologically separate; then they are reunited through some sort of connection, such as a "network," but even when this is done, it is clear that the entities, now found to be related, preexist the connection.
>
> (2020: xiv)

Weak relationality "stem from ontologies that are deeply embedded in the negation of the full humanity of multiple others and the nonhuman" (ibid). In strong relationality, "nothing preexists the relations that constitute it; in other words, reality is relational through and through" and it stems from "nondualist ontologies and their corresponding pluriversal forms of politics" (ibid). Strong relationality in pluriversal politics embodies not only non-dualist ontology but also "weak ontology" as cultivated by Giani Vattimo (1999). Vattimo makes a distinction between "weak ontology" and the "strong ontology." Strong ontology as in Heidegger's early phase is one of mastery while weak ontology is one of vulnerability and mutual care and responsibility. Escobar's strong relationality and political ontology embody weak ontology as well as a weak naturalism as discussed by Habermas (2003) where we work with Nature but are not determined by it. Weak naturalism embodies our creativity, courage, and immanent transcendence. In the face of the current ecological holocaust, we need to cultivate a different political ontology of weak ontology, weak naturalism, and strong relationality which would help us go beyond the current challenges of the climate and social collapse and embody responsibility. As Escobar (2020: 25) writes:

Political ontology refers in the first instance to the practices involved in creating a particular world or ontology; it also provides a space for studying the relationships between worlds, including the conflicts that result when different ontologies or worlds strive to preserve their existence in their interactions with other worlds, under asymmetric conditions of power. Political ontology exists in the space between critical currents in the academy and the ongoing struggles to defend territories and worlds. It reveals the ontological dimension of accumulation through dispossession that is taking place in many parts of the world under extractivist development models, especially in large scale mining, biofuel production, and the appropriation of land linked to commercial agriculture. It lets us see why environmental conflicts are often at the same time ontological conflicts—that is, conflicts over contrasting ways of existing and making worlds. Finally, political ontology seeks to highlight and promote the pluriverse while resisting the tendency to represent the world as if it were only one. It records the rise and political mobilization of relationality as a space for struggle and life force. It bears witness to the urge to rebel among many communities faced with the ravages caused by a world that has arrogated to itself the right to be "the world."

To this field of conversations, we can bring the insights of Piet Strydom, a thoughtful social thinker of our times. In his reflections on Climate Change and the Anthropocene, Piet Strydom (2015) urges us to realize the significance of "cognitive fluidity." Cognitive fluidity reflects the capacity of mind and cognition to creatively respond to the challenges of both climate and society without being a prisoner of it. It reflects weak naturalism where we work with Nature but are not determined by it. For Strydom, cognitive fluidity helped us to come to terms with climate change 10,000 years ago which was accompanied by the birth of agriculture and formation of sedentary forms of life. In our current age of climate change and Anthropocene, we need to creatively embody cognitive fluidity and unblock our processes of societal, cultural, ethical, and aesthetic learning.[5] We also need to engage in critique of contemporary structures of economy, polity, science, self, and society which has given rise to pathological forms of cognition and social organization leading to our imminent climate and social collapse. Strydom calls this explanatory critique.[6] For Strydom, embodying cognitive fluidity, weak naturalistic ontology, explanatory critique, and creativity which is transdisciplinary going beyond both scientism and naturalism, sociologism and culturalism help us cultivate visions and practices of responsibility as part of an appropriate subject-formation as well as formation of creative institutions for our new planetary society.

To these reflections, Sailesh Rao, originally from India and now based in the USA and working on climate change, brings a spiritual approach to this. He uses the languages of *Dharma*—right conduct—and *Yoga*—to come to terms with the challenge of carbon emission and climate change. In his book *Carbon Dharma: The Occupation of Butterflies*, Rao (2016a) writes how we have do the dharma of butterflies now connecting to both Nature outside and inner nature which also questions the premises of our current binary civilization dividing between inner and outer. For Rao, we have to be aware of our predicament and roles as caterpillars trapped by many fears of present, past, and future and do our dharma as butterflies embodying the dharma of compassion for self, other, Nature, and the World. As Rao tells us in his *Carbon Dharma*:

> Reconnecting humans back to Nature is about reconnecting humans back to reality by identifying and overturning all the absurd notions that underlie our current civilization. It is also about reconnecting humans back to our own selves. Nature isn't just out there in the wilderness, Nature is within us as well. Put it it another way, to dissolve the illusion that we are separate from one another and Nature, it is necessary to dissolve our ego which promotes our illusion of separateness and thus connect within to the Atman, the Spirit within us. To become compassionate towards all Life and thus fulfill our purpose as human Butterflies, we must practice to be self-compassionate. As such the Metamorphosis is a deeply spiritual undertaking.
>
> (Rao 2016a: 21)

For Rao, as in Nature, so in life and society, caterpillars destroys and eats all that it surveys while the butterflies comes out of the Pupa and "pollinates the flowers helping to regenerate Life" (Rao 2016a: 9). Chakraborty's earlier call for *Homo sapiens* to be *Homo prudens* gets a creative new vision and pathway of realization in Rao's paths of becoming butterflies, as he writes:

> As a species, Homo Sapiens, Latin for "Wise Man," is most definitely in its Caterpillar stage of development. It is a voracious consumer, a waste-producing eating machine that has munched through most of the complex Life in the ocean and half the forests on land. As a species, I believe that it is well past due for Homo Sapiens to grow into its pupa stage, emerge as a Butterfly and begin udoing the destruction wrought earlier.
>
> (Rao 2016a: 1)

In his book, Rao gives us examples of human butterflies who work with abandoned land and people and make them blossom. "Butterflies heal, while Caterpillars destroy" (2016a: 11). Rao tells us:

> It is the identification with the ego that drives the Caterpillar side of a human being. The Caterpillar is driven by the fear that arises out of sense of sepration and isolation. It is the identification with the Atman

[Soul, Self] that drives the Butterfly side of a human being. The Butterfly is driven by the love that arises out of a sense of belonging.

(2016a: 21)

In his other book, *Carbon Yoga: The Vegan Metamorphosis*, Rao tells us, "We need compassion as much as we need air. . . . Compassion for all Creation is infinitely sustainable. Conversely, violence to any part of Creation is unsustainable" (Rao 2016b: 55). For Rao, "Compassion for all Creation or kindness to all Life is summarized in a single, ancient Sanskrit word, Ahimsa" (ibid: 54). Climate change and our contemporary condition of Anthropocene call for a yoga of Ahimsa, a yoga of active non-violence. In our chapter on global justice and responsibility, we have met pioneering activists such as P.V. Rajagopal of Ekta Parishad who put forward Ahimsa as a path and goal of global justice and responsibility. Similarly, Ahimsa—active non-violence—is a path and goal of climate justice. In fact, it is Ahimsa which can connect threads of concerns in Chakraborty, Latour, Haraway, Lovelock, Escober, and Strydom and cultivate creative responses to climate change and the Anthropocene. As political philosopher Ramin Jahanbegloo tells us building on Mahatma Gandhi and Martin Luther King, Jr. which also resonates with the thoughts of thinkers such as Judith Butler: "Living is a matter of nonviolent organization of the world, with the aim of becoming more empathetic and interconnected" (Jahanbegloo 2020: 26).

Big History has been another important intellectual and consciousness movement of our times which urges us to realize our interdependence with other species and times. Pioneers and participants associated with this movement such as David Christian, Barry Rodrigue, and others link the current discourse of the Anthropocene to the need for greater collaboration and evolution (see Christian 2004; Rodrigue et al. 2017). They invite us to understand our past and present from the Big Bank to the present which calls for a multi-temporal hermeneutics in the sense we have discussed this in the previous section, that is, we need to move from the Big Bank to the present and vice versa. In the last Annual Conference of International Association of Big History, "Changing the World: Community, Science, and Engagement with Big History," August 1–4, 2021, held virtually in collaboration with Symbiosis School of Liberal Arts, Pune, some of the key players in the field of Big History such as David Christian strove to make links between Big History approach and the discourse of the Anthropocene. For Christian, in the Anthropocene, we are interdependent on the global scale. Emlyn Koster, a participant of the Conference, told in the conference that humanity has become estranged from the planet. Koster invites all of us develop ecological empathy and conscience in our age of Anthropocene which would help us overcome our current challenges of climate crises and social crises (see Koster 2019).

Refugees and Responses: Challenges of Responsibilization

War and terrorism such as in Syria and Middle East have created problems of refugees. Poverty and deprivation in Africa and around the world have also created a great refugee problem. Different countries in Europe have offered differential responses to the refugee crises. Countries such as Germany have shown a great deal of maturity in dealing with these crises. Germany has welcomed millions of refugees from Syrican crises. On the other hand, the USA with the election of President Donald Trump has been harsh on the refugees to put it mildly and was building a wall across Mexican border. With the change of regime in the USA, with President Joe Biden and Vice-President Kamala Harris, there seems to be softening of visibly cruel approaches to the refugees. The refugee crises call for creative responses and movements of global responsibility. We have seen how Indignados movement is concerned about this in Spain.

University Social Responsibility (USR)

Recently some universities around the world such as University of East Anglia in the U.K. and University of Hyderabad in India have come together to create initiatives in University Social Responsibility. The effort is to make universities more socially responsible, which goes beyond looking at universities as functional units and corporations and establishing creative and responsible relationship with society.[7] Professor Prabhakar Jandhyala from University of Hyderabad in this regard tells us that University of Hyderabad has been partnering with voluntary organizations for delivering services to COVID-19 patients and trying to take up more socially relevant issues. Professor Daniel Rycroft of University of East Anglia here brings other related wider issues to the discourse. As a scholar of art, Rycroft suggests how USR can promote more aesthetic activities within, across, and beyond university spaces. He also argues that USR involves much more inter-institutional and inter-sectoral collaboration within and beyond university which goes beyond the existing logic of interdisciplinarity. As part of an interlinked global vision and practice, USR can link to UN sustainable development goals and empower citizens to in the lives of universities and society more meaningfully and also realize their responsibility as global citizens—citizens of the world. Given contestations about citizenship and refugee crises in India and around the world on the wake of the promulgation of CAA (Citizenship Amendment Act) in India in December 2019, universities can help realize humane aspects of citizenship beyond bounded nationality and legality as suggested in the works of Ambedkar, Gandhi, Tagore, and Sri Aurobindo from India and similar transformative figures from around the world. Rycroft here also argues that USR does not stand alone and it is part of a related field of responsibility consisting of Intellectual Social Responsibility, CSR, Academic Social Responsibility, and Human Responsibility. Among these, Academic Social Responsibility and Intellectual Social Responsibility have the potentiality to involve much more critical and creative perspectives and engagement.

COVID-19 and the Challenges of Trauma and Responsibility

> The viral epidemic is basically a matter of health and disease. But there is another epidemic of a public or civic kind which reduces politics basically to the conflict between friends and enemies. This conception has tended to become dominant in recent times world-wide—and this is why it can also be called a pandemic. This conception is opposed to a view which sees politics mainly as the struggle for the "good life" of all. The key term of the first conception is "power", the key term of the second view is "justice". The two kinds of pandemics are different: the viral pandemic attacks or destroys the physical body, the civic pandemic attacks and destroys the public body or the spirit of civic life. One needs to keep them apart. However, there are also overlaps or instances where the two epidemics become fused. This happens when the politics of "power" seeks to control the pandemic of health, that is, where political rulers seek to manipulate the health epidemic.
>
> —Fred Dallmayr (personal communication)

> This was a pandemic that was described and reported in terms of statistics—numbers of infections, numbers of patients in critical care and numbers of death. Lives were transformed into mathematical summaries.

> But those who died must not be summarized. They must not become lines on squared paper. They must not become mere rates used to argue differences between nations. Every death counts. A person who died in Wuhan is as important as one who died in New York. Our way of describing the impact of the pandemic erased the biographies of the dead. The science and politics of COVID-19 became exercises in radical dehumanization.
>
> Richard Horton (2020), *The COVID-19 Catastrophe:*
> *What's Gone Wrong and How to Stop It Happening Again*

COVID-19 is one of the most serious challenges before humanity today and different societies are facing it differently and it heightens the need for deepening our visions and practices of global responsibility. To understand this, we need to understand our contemporary human social condition. With the spread of the Corona virus and the rising death and destruction, our contemporary COVID-19 condition challenges us to understand the different dimensions of it. With Corona virus, it is not only a case of viral pandemic but also one of civic pandemic (Horton 2020; Fang and Berry 2020).[8] This civic pandemic is manifested by the use of authoritarian means to deal with pandemic, as it happens with handling of this disease in some countries. Corona virus leads to the damage of our respiratory system and eventually our inability to breathe. But here historian and philosopher Achille Mbembe (2020) challenges us to realize that humanity was already threatened with suffocation before the Corona virus.[9] This becomes evident with the murder of George Floyd in Minnesota, the USA, in the hands of the police officers whose last words were, "I cannot breathe." This murder led to wide-spread protests and movements such as Black Lives Matter and movements for police reform in the USA. This also reminds us the challenges of conjoint fight against virus, racism, and endemic poverty as carried by leaders such as Jane Addams in the USA who fought against poverty, racism, and the Spanish flue. For Mbembe, Corona virus is also related to our problems of co-living such as racial co-living but also living with other species reminding us of the works of pioneers such as Donna Haraway whose work we discussed in our previous sections.

Towards a Critical Genealogy and Ontology of Our Corona Presents

It is our inability to live with respect and concern with other species and our mindless and unconcerned destruction of forests and habitat that has led to the unleashing and transmission of viruses such as Corona virus which also puts questions to the working of contemporary human civilization.[10] Thus, the current Corona crisis is related to crisis of civilization (Chakraborty, Dipesh 2020). It is also related to crises of climate and capitalism, as critically suggested and argued by activists and scholars such as Greta Thunberg (2019), Bruno Latour (2020), and Slavos Zizek (2020). Understanding our contemporary Corona condition challenges us to understand what Michel Foucault (1984) calls a critical ontology of the present which is historical as well as animated by an urge to overcome the fatalism of the present and create alternative different presents and futures. Such an approach is also facilitated by Giani Vattimo's (2011) pathway of ontology of actuality which can also be realized as ontology of actualization. At the heart of Vattimo's project is the work of weak ontology which is a realization of our own limits and rather than asserting our own knowledge, ignorance, arrogance, and power in a strong way. Our contemporary Corona condition brings to the fore our inability to know our own uncertainty, as many critical thinkers such as Jurgen Habermas (2020) and Veena Das (2020a, 2020b, 2020c) have challenged us to realize.[11] The challenge before us is to fashion appropriate public policies of containment and healing based upon our ignorance and limited knowledge rather than what economist Jishnu Das (2020) calls an "epidemic of ignorance."[12] Acknowledging our limits and uncertainty calls for a new practice of knowing, a more humble as well as courageous way of knowing, a "new epistemic humility,"[13] new border crossing between ontology and epistemology which can be called an ontological epistemology of participation (Giri 2006). Our contemporary COVID-19 condition calls for an appropriate ontological epistemology of participation.

COVID-19 and the Challenges of Trauma, Solidarity, and Responsibility

Corona virus creates trauma which is both natural and social. But while this threat to humanity should have led to greater collaboration among nations and peoples, it has led to avoidable conflicts and struggles for power. A case in point is the geopolitical struggle between the USA and China during the spread of the Corona pandemic. Another traumatic part of our contemporary Corona moment is the aggression of China. Instead of doing all it could do to help humanity deal with these crises, China started aggression against countries like India in the Galwan valley in Ladakh as well as threatening countries such as Vietnam and a show of strength in South China Sea. Such aggression is accompanied by manifest authoritarianism in the decision of some elected political leaders which is adding injury to the salt. Thus, the trauma of the virus is multiple as we are confronted not only with the virus and the vaccine but also with the challenge of veracity as sociologist Jenny Reardon (2020) challenges us to realize. Here we need to realize the manifold interlinked challenges of virus, vaccine, veracity, and victory where we strive to realize victory not only over virus and Nature but also with virus and Nature. This also calls for appropriate construction of trauma in which awakened individuals and social movements play an important role which can build on earlier works on social constructions of risk (see Beck 1992; Strydom 1999). Only with such creative articulation and construction, trauma can be transformed into responsibility as suggested by the important work of cultural sociologist Jeffrey Alexander and his colleagues (Alexander et al. 2004). Both in India and the USA, we see such critical and creative construction at work *vis-a-vis* raising voices against both racism as well as the authoritarian governance of the COVID-19 pandemic. In the Indian context, journalists and human rights activists played an important role in bringing to light the immense suffering of the migrants walking back home or boarding the migrant express and some dying in the train (Hans et al. 2021; Mander 2020). Such construction of trauma calls for a new relationship with reality and constructiveness. As Greta Thunberg (2019) suggests in the context of the related challenge of climate crises, we need to think outside the box for this which would help us cultivate what she calls "cathedral thinking." And here Judith Butler (2020) challenges us to realize that for such critical and creative constructiveness where we work with the challenge of the pandemic to produce solidarity, we have an ethical obligation to be unreal (see Butler 2020; Gessen 2020; Weir 2020).

In this creative act of constructivism, writing plays an important role as well as gives us strength to go beyond the fatalism of the present. As Fang Fang writes in her much-talked *Wuhan Diary* about the Wuhan lock down in China:

> Since most of the residents in Wuhan do not have their own automobiles, they had to walk from one hospital to another in search of a place that

might admit them. It is hard to describe how difficult that must have been for those poor patients. . . . We all felt completely helpless in face of these patients crying out, desperate for help. Those were also the most difficult days for me to get through. All I could do was write, so I just kept writing and writing; it became only form of psychological release.

(Fang and Berry 2020)

Writing here is linked to a new vocation of being and becoming and it is also linked to a new art of responsibility (Das 2020; Giri 2020; Momaday 2020). To respond to challenges of COVID-19 as well as the related challenges of climate change, we need creative and critical visions and practices of responsibility. In our present-day world, we predominantly move in frames of rights and justice, but these are not adequate to deal with our challenges and we need creative frames, institutions and practices of responsibility at the levels of self, society, state, and our international order (Strydom 2000). While COVID-19 threatens the whole human world, what needs more conscious cultivation is the corresponding movement of global responsibility. In this context, Ramin Jahanbegloo (2020) in his recent book on Coronavirus tells us:

In the tragic and increasingly exacting battle that is taking place between humanity and its new enemy, COVID-19, the common suffering of human beings has paved the way for a broader solidarity of all individuals across all lands. A global march like this on a long and dark road is a new endeavor for humanity, perhaps the most significant since the fight against Nazim in the 1940s.

(2020: 59)

In his chapter, "Solidarity of the Shaken," Jahanbegloo tells us:

We can deal with global tragedies, like the coronavirus outbreak, only if we learn to change our modes of being-in-the world and doing-in-the world. This crucial and critical renewal consists above all in always substituting responsibilities for rights. Being obsessed with rights alone means dismissing the dimension of solidarity as the art of forging a unity with others. The coronavirus outbreak has demonstrated that the fundamental right to live is suspended from the duty of each person to respect the instructions of containment. Therefore, developing a sense of solidarity with Others as part of a global citizenry is paramount as each of us is dependent on others. Tracking the question of solidarity in real life, one needs the idea of responsibility.[14]

In this call of "Solidarity of the Shaken," we have to realize the urgency of the solidarity with the shaken. This is particularly true of migrant workers who had to leave their places of work and then to come back home walking of travelling in most precarious circumstances. The Prime Minister of India announced a country-wide lockdown at four-hour notice in March

2020 which led to immense suffering and miseries. Many people died out of starvation and exhaustion while travelling or walking back home. In his essay on this theme, Manoranjan Mohanty raises the question of rights, responsibility, and fundamental freedoms that the plight of migrant workers poses. Mohanty urges us to realize here:

> The real challenge that came out boldly during the COVID-19 crisis was the need to ensure everyone's right to work with dignity in their home regions. Much of migration is distress migration. Therefore, the prevailing economic process must be re-examined so that people get work in their own village or town or nearby area and do not migrate. Thus, restructuring the political economy to facilitate local employment and local development is the urgent need. Once an area is developed, the migrant labour, more as mobile labour, can go with a higher bargaining power and adequate facilities to help meet the labour demand in certain areas of the country or abroad. We need to initiate restructuring of the rural economy as a whole so as to provide long-term solutions to poverty and unemployment. Rather than aiming at a five trillion GDP, achieving full employment should be our goal. . . . MGNREGS can be re-imagined to cover the entire rural economy rather than specific types of jobs listed in a schedule. A diversified rural economy combining traditional and modern technologies can be planned by the panchayats as a zero-unemployment development strategy. The prevailing system of neoliberal, growth-centric economic model steered from above which throws crumbs as relief to the poor under various programmes needs to be transformed into a decentralised self-propelling, sustainable development process at the grass-roots level that makes it possible for people to realise their fundamental rights to live with.
>
> (2021: 5–6)

Engaging with the precarity of the migrant labourers and the need for an alternative politics which recognizes the significance of labour, Mohanty further tells us:

> In my view, the alternative is a new politics of labour that fundamentally challenges the political economy of gradation and degradation of labour and economies. It must reject the charity-and welfare approach of capital and state and affirm the rights approach instead, thus moving from an instrumentalist view to a substantive view of recognising labour as the principal force of civilisational development.
>
> (2021: 9)

The Challenges of Transformations

The current trauma of the pandemic calls for multi-dimensional transformations. For example, it calls for transformation of nationalistic jingoism including vaccine nationalism where nation-states with more resources and vaccine capacity are not always sharing this with less resourced and capable people and countries. This calls for greater co-operation among nations, individuals, and cultures. It also calls for transformation of the nation-state. As Arjun Appadurai (2020) argues, the pandemic does not kill globalization but calls for a different kind of globalization and transformation with transformation of the logic of the nation-state. In line with our earlier discussion of Chakraborty's distinction between the global and the planetary, the pandemic urges us to transform the global mode to the planetary. It also calls for building new transnational institutions and movements. It also calls for building new regional movements and institutions for economic, political, and climate-related collaboration. The current logic of long-distance production and consumption which is a direct offshoot of our contemporary economic system is not sustainable. The pandemic challenges us to build new ecologies of production and consumption. Here social activist Ela Bhatt (2015) challenges us to build 100-mile communities where producers and consumers would be able to exchange their production and consumption.[15] And here anthropologist and creative thinker Alf Hornborg argues that we need to have local currencies in our interactive communities of trade and exchange which would help us in localization as well as cross-local regional and planetary collaboration (Hornborg 2017, 2019).

The current COVID-19 pandemic calls for creative transcendence of existing borders and boundaries such as nation-state-erected borders. It also calls for a new art of border crossing where we can overcome our existing epistemic and ontological borders and cultivate new ways of learning and being together. It also calls for new pathways of cross-cultural dialogues so that we can realize new meanings of death, disease, well-being, life, and healing.

The pandemic also calls for political transformation, a new kind of politics. It calls for more collaborative leadership rather than single person or authoritarian leadership (see Willis 2021). We see that such leadership which listens to many voices has been able to stem the spread of the pandemic. Also, the countries that are led by women leaders such as Taiwan, New Zealand, Germany, and Finland have been able to control the spread of the virus. This has led to a very important discussion about the differential significance of women leaders in providing creative and collaborative leadership in handling critical crises such as the present pandemic. We need empathic leaders who can feel the pain and suffering of self and others and listen to them, who can work as "apostles of ear," who become apostolic in their visions and practices of listening.[16] We also need leaders who can listen to the voices of their souls and listen to the rhythms of their bioregions—biological diversity as well as cultural diversity—and the whispers and groans of our Mother Earth (see Howard et al. 2019). We need leaders who become bio-regional as well as Soul-Planetary or *Atmic* Planetary as suggested in our previous section on corporate social and spiritual responsibility.

A related challenge is also practicing a new kind of politics what Arturo Escobar (2020) calls pluriversal politics which gives spaces and voices to the many rather than the one-dimensional nation-state-centred system of politics of modernity. We find intimations of a new kind of politics from contemporary critical scholars such as William Connolly (2013), Donna Haraway (2016), Judith Butler (2020), and P.V. Rajagopal (see Reubke 2020).[17] They challenge us to cultivate a new political imagination and practice which is less violent and based upon collaborative self-organization and mutual organization. Such a politics would rethink the current project of rights and justice and link it to responsibility. While working on justice, it would acknowledge the significance of interpretation in realizing justice (Dworkin 2013). Justice here is also not confined among human beings; much more attention is given to the art of living together justly with other beings, especially other species what Martha Nussbaum (2006) calls "cross-species dignity." The interpretative exercises of justice are here also not anthropocentric or shrouded in the veil of ignorance and arrogance of nation-state; rather these involve sympathetic and radical movements across borders where we interpret life and justice not only from the primacy of the human and state-centred and capital-centred rationality but also from the points of view of all lives. This way justice and responsibility become multi-*topial* where we move across different *topoi* and terrains and strive to create livable worlds and Earth by putting ourselves in the feet and trails of other species (Giri 2018a).

Ethical Issues

These issues of political transformations and a new kind of politics are related to issues of ethics. To live with and beyond the pandemic, we need to live ethically and here we can build upon multiple understanding of ethics and morality and multiple traditions of it. In this context, along with many contemporary approaches, the project of ordinary ethics as cultivated by Veena Das (2020b, 2020c, 2015), Michel Lambek (2010), and others is helpful where we live ethically in our everyday lives acknowledging our vulnerability and, at the same time, realizing our capacity to resist degradation and create new possibilities.[18] A related project is a project of emergent ethics where we work on emerging ethical sensibility and norms in our contingent situations and our location in multiple contingencies of self, other, state, market, social movements, and the world (cf. Quarles van Ufford and Giri 2003). Such an emergent ethics is different from an emergency ethics which is imprisoned within a logic of state of exception and based upon fear and terror (see Agamben 1995). Another project is a project of aesthetic ethics where ethics creates an art of living—an ethos of artistic living—building upon both ethics and aesthetics (Ankersmit 1996; Clammer and Giri 2017). Our current COVID-19 condition calls for creative works and meditations of aesthetic ethics in our lives which also involves critiques of egoistic and possessive individualism and affirmation of "social and ecological interdependence, which is largely misrecognized as well" (Butler 2020).

The Spiritual Calling of COVID-19

Our contemporary condition also brings us to the spiritual dimension of our existence as well the *sadhana* and struggles for transformations. Spirituality is confined not only to our individual well-being but also to our collaborative well-being as suggested in Martin Luther King's vision and practice of Beloved Community (King 1967). Spirituality helps us in our strivings and struggles to live with beauty, dignity, and dialogues (Giri 2018b). It also challenges us to realize the agonies of our life and then engage in strivings and struggles to transform it thus becoming agonal spirituality. Such an agonal spirituality which resonates with the project of agonal democracy as suggested by Ernesto Laclau and Chantal Mouffe (1985) helps us to realize spirituality as both compassion and confrontation. The rise of authoritarianism and fundamentalism of many kinds calls for confrontation as well as compassion. Here we can build upon spiritual possibilities in our religious traditions. With regard to the COVID-19 pandemic, we can realize that many religious and spiritual traditions urge us to wear masks such as the Jaina traditions. The Jaina tradition urges us to relate to the virus as an entity and not just as an enemy challenging us to beyond the dominant enemy trope of modernistic politics, political theology as well as our current war on virus strategy (see Fang and Berry 2020). The Jaina tradition has a practice of dying peacefully and with our voluntary effort and in the context of the current pandemic of death and destruction, we need to learn how to die and also to live with dignity and wisdom (Chapple 2020). This is also the spirit of a philosophical approach to our current situation as suggested by the noted philosophers Simon Critchley (2020) and Ramin Jahanbegloo (2020).[19]

COVID-9 calls for a new common sense. COVID-19 and other pandemics are related to our inability to hear the voices of Nature, one aspect of our life with senses is to cultivate our hearing senses so that we can listen to the cries of both Nature as well as vulnerable social groups such as the migrant workers whose lives were given a toss by the powers that be by announcing a lock down at a four-hour notice. Primal people of the world have far more developed sense to listen to each other and Nature. But our modern social sciences privilege speaking over listening. For realizing further possibilities with our senses such as listening, hearing, and seeing, we need to further develop our senses which would contribute to a deeper realization of common sense.[20] Common sense is also not bound only to the realm of sense perception, that too empirical sense perception. It also touches what Pitrim Sorokin (1985) long time ago called the super-sensate.

Corona virus is creating death and suffering and much of it is avoidable if we have the right public policy as well as right ways of living at the levels of individuals, families, and community. At the same time, the current pandemic challenges us to live differently and live with death differently and resist what Horton (2020) calls "radical dehumanization."[21] Modernity has been primarily preoccupied with life, that too young and successful life, and it has put death into background. We need to learn to live with death creatively and meaningfully. We need to have dialogues with death. In this dialogue with death, we can learn from multiple traditions of humanity. In the Indic traditions, there are insightful dialogues with Death. In the dialogue between Yama and Nachiketa in Kathopanishad, Nachiketa is not afraid of death and wants to use the opportunity of dialogue to realize the meaning of life and immortality. This theme of immortality again arises in the famous dialogue between Yajnavalkya and Maitryeyi in *The Brihadaranyaka Upanishad*. Yanjnyavalkya tells Maitryee that the world's wealth cannot give her immortality and her life would be as miserable as life of other rich people (see Sen 1999). There is an important dialogue between Yaksha and Yudhisthira in Mahabharata where this dialogue is happening in the context of death of his four brothers who rushed to drink the water in the lake without listening to the voice of Yaksha to answer his question. This was an act of arrogance which is related to arrogance of many of us in our contemporary moment of the pandemic. Yaksha asks Yudhisthira who is a living corpse, and Yudhisthira says, "He who does not worship the following five—'Gods, guests, family, dependents and soul'—is a living corpse though living normally" (Murthy 2004). There is also dialogue between Savitri and the King of Death in the Mahabharata as well as in Sri Aurobindo's epic *Savitri* in which Savitri is striving to revive her dead husband Satyavan. This can be read and realized together with the scene of a young boy in a railway station in Bihar where he is trying to awaken her sleeping mother without him not knowing that her mother is dead after an exhaustive journey back home in a COVID-special train during the lockdown in India. During this pandemic, many people—front line health workers as well as ordinary human beings—are trying to put their lives in risk and revive others. This is a creative and courageous act which challenges us to create life in places of death and destruction. Through such creativity and courage which is shown by many people in saving and nurturing lives during our pandemic we overcome mortality and cultivate immortality.[22] The current pandemic brings us to both crises of life as well as death as part of our current civilizational crises. It challenges us to cultivate a new civilization of life and death.

Both science and spirituality challenge us to realize that the virus is not just our enemy. Human life has evolved with the virus (Tanabe 2020). But the dominant discourse is a war on virus. A spiritual engagement tells us how we can embrace the challenge of the virus in a new way, do a yoga with Corona virus which can become viral. We do a new Corona yoga which leads to a yoga of *karuna*-compassion. We just cannot win victory over the virus and Nature but we win victory with virus and Nature as our intertwined story, *sadhana*, and struggles for creating livable worlds of beauty, dignity, and dialogues for all.

COVID-19 and Alternative Futures

In the context of our current predicament, there is a discourse of the new normal. But the new normal with lock down and many other aspects of our contemporary condition such as authoritarianism and violence is also pathological. It is in this context that we need to rethink our present and pathways to futures critically and creatively. We need to work with and transform both our new normality and pathology and realize as Axel Honneth (2007: 35) argues, "A paradigm of social normality must, therefore, consist in culturally independent conditions that allow a society's members to experience undistorted self-realization." Honneth continues, "The question then comes crucial whether it is a communitarian form of ethical life, a distance-creating public sphere, non-alienated labour, or a mimetic interaction with nature that *enables individuals to lead a well-lived life*" (ibid). In their recent work on migration and COVID-19, Asha Hans et al. also tell us:

> It is time to start the pedagogic imagining and structuring of a future world that will open up to new possibilities. The questions before us are: what is normal and just, and how do we protect our fundamental rights when these rights are trampled on? In this context, questions we should be then asking are: what is the appropriate language to create a new alternative? How do we work in collaborative ways? We also need to ask: how do we stop this violence from becoming the "new normal" in our lives? Are we prepared to re-imagine new worlds where security is dependent not on force but recognition of an interdependent world of collaboration? It is imperative to introduce different methods of development and a new alternative to the "new normal". Would this mean creating fundamental changes in our understanding of words such as "poverty"? This new manifestation is the shift that strikes at the very core of our social structure calling for a reconceptualisation of the vocabulary of change? We recognise that the pandemic has essentially affected our lives and our humanity and is the catalyst we hope for a new equal and just global system.
>
> (Hans et al. 2021: 5–6)

In this context, we need to transform our suffering embedded in these questions and our implicated existence into healing—self, social, and global. As Memembe (2020) here challenges us to realize:

> In the aftermath of this calamity there is a danger that rather than offering sanctuary to all living species, sadly the world will enter a new period of tension and *brutality*. In terms of geopolitics, the logic of power and might will continue to dominate. For lack of a common infrastructure, a vicious partitioning of the globe will intensify, and the dividing lines will become even more entrenched. Many states will seek to fortify their

borders in the hope of protecting themselves from the outside. They will also seek to conceal the constitutive violence that they continue to habitually direct at the most vulnerable. Life behind screens and in gated communities will become the norm.

In the context of our current predicament, there are varieties of talks about post-COVID futures. But without multi-dimensional transformations—social, economic, political, and spiritual—our post-COVID futures may not be different from our current condition. In this context, we need to cultivate alternative planetary futures both in discourse and in practice. But future is not only a fact—a cultural fact—but also a matter of values (Appadurai 2013). We are challenged to create pathways of beauty, dignity, and dialogues and alternative planetary futures which are not reproductions of existing dead and killing systems and ways of thinking. As Arundhati Roy (2020) challenges us to realize in her challenging reflections, "The Pandemic is a Portal:"

> What is this thing that has happened to us? It's a virus, yes. In and of itself it holds no moral brief. But it is definitely more than a virus. Some believe it's God's way of bringing us to our senses. Others that it's a Chinese conspiracy to take over the world.
>
> Whatever it is, coronavirus has made the mighty kneel and brought the world to a halt like nothing else could. Our minds are still racing back and forth, longing for a return to "normality," trying to stitch our future to our past and refusing to acknowledge the rupture. But the rupture exists. And in the midst of this terrible despair, it offers us a chance to rethink the doomsday machine we have built for ourselves. Nothing could be worse than a return to normality. Historically, pandemics have forced humans to break with the past and imagine their world anew. This one is no different. It is a portal, a gateway between one world and the next.
>
> We can choose to walk through it, dragging the carcasses of our prejudice and hatred, our avarice, our data banks and dead ideas, our dead rivers and smoky skies behind us. Or we can walk through lightly, with little luggage, ready to imagine another world. And ready to fight for it.

The Calling of Global Responsibility

Epistemicide, Epistemic Freedom, and Epistemic Responsibility

Realization of global responsibility is also confronted with the challenges of overcoming epistemic and cognitive injustice and realizing epistemic freedom and responsibility. Our current knowledge space is full of epistemic violence and violence animated by Euro-American domination leading to what Boaventura de Sousa Santos (2014) calls epistemicide. Epistemicide is accompanied by Terricide or killing of Earth[23] In this context, de Sousa Santos (2014) calls for going beyond the domination of Northern epistemologies and bringing epistemological paths and perspectives from the South. Here de Sousa Santos goes beyond a facile dualism between South and North (see Giri 2018a). This work today can be part of conversations across borders and planetary conversations. In this context, Sabelo J Ndlovu-Gatsheni (2018) speaks about realizing epistemic freedom in Africa which has a global relevance.

In the introduction to his book, *Epistemic Freedom in Africa*, "Seek ye Epistemic Freedom First," Ndlovu-Gatsheni challenges us to realize that epistemic freedom is not just academic freedom understood conventionally:

> Epistemic freedom is different from academic freedom. Academic freedom speaks to institutional autonomy of universities and rights to express diverse ideas including those critical of authorities and political leaders. Epistemic freedom is much broader and deeper. It speaks to cognitive justice; it draws our attention to the content of what it is that we are free to express and on whose terms. . . . Epistemic freedom is about democratizing "knowledge" from its current rendition in the singular into its plural known as "knowledges." It is also ranged against over-representation of Eurocentric thought in knowledge, social theory, and education. Epistemic freedom is foundational in the broader decolonization struggle because it enables the emergence of the necessary critical decolonial consciousness.
>
> (ibid: 4)

He also tells us, "What is emerging is the importance of epistemic freedom as the foundation of other freedoms. Epistemic freedom has the potential to create new political consciousness and new economic thought necessary for creating African futures" (ibid: 80).

Struggle for epistemic freedom begins with the realization of the violence and humiliation of the epistemic line which puts seekers from outside the Euro-American epistemological zone into a position of permanent inferiority. For Ndlovu-Gatsheni,

> If the "color line" was indeed the major problem of the twentieth century as articulated by William E. B. Du Bois [. . .], then that of the twenty-first century is the "epistemic line." The epistemic line cascading from the "color line" because denial of humanity automatically disqualified one from epistemic virtue.
>
> (p. 3)

But Africa has long genealogies of knowledge traditions, a "long history that predated its encounter with Europe" (p. 2).

In his reflection on epistemic freedom, Ndlovu-Gatsheni tells us about the significance of "onto-decolonial turn," which creatively challenges us to bring both the ontological and decolonial turn together: "The present age of anthropocene is simultaneously driven by the 'ontological' and 'decolonial turn' turns constituting what can be named as 'onto-decolonial turn'" (p. 71). In his epilogue to this volume, Ndlovu-Gatsheni also emphasizes the significance of de-colonial turn for realization of global responsibility and offers "Ten-Ds" of decolonial turn—Deimperialization, Desecularization, Depatriachization, Debourgeouisement, Decorproatization, Democratization, Decanonization, Deborderization, and Deanthropocentrism (see the Afterword section). This is an important critical initiative as it can interrogate other important contemporary turns such as feminist turns, ecological turns, and linguistic turns which, according to R. Sundara Rajan (1998), the creative critical philosopher from India, define important turns in discourse and practice in our contemporary world. All these turns, as important as they are, suffer from the one-sided emphasis of the epistemic and the neglect of the ontological. The feminist turn, for example, is primarily epistemic and lacks projects of ontological co-nurturance of both the male and the female as complexly constituted separate but related beings. It was also not sufficiently attentive to the questions of colonialism to begin with. But these turns themselves now can be interrogated and transformed with onto-decolonial frames and movements of critiques, creativity, and transformations. But onto-decolonial turn also needs a further turning and tuning what can be called a spiritual turn. Following contemporary writer and critic Pankaj Mishra, Ndlovu-Gatsheni tells us how this turn emerged "within a context and age of anger and hate" (ibid: 71). But the challenge is how to overcome hatred and anger. The onto-decolonial turn needs to take a spiritual turn in terms of overcoming temptation to anger and retaliation and establishing compassion and solidarity. As Ndlovu-Gatsheni himself writes, "What emerges poignantly here is that a decolonial reconstitution of the political has to fundamentally redefine humanity in non-separatist terms and inaugurate a new dispensation of decolonial love" (ibid: 93). Realization of this calls for transformation of self, society, politics, religion, and spirituality. Spiritual turn points to such a multi-dimensional movement of transformation which is simultaneously practical and spiritual. It is a multi-dimensional striving and struggle for realization of beauty, dignity, and dialogues in self, cultures, societies, world, and cosmos which is part of movements of both practical discourse (in the sense of Jurgen Habermas's articulation of this as a project of realization of moral argumentation and democratic transformations) and practical spirituality.

In the concluding chapter on "African futures," Ndlovu-Gatsheni links the struggles for epistemic freedom in Africa to alternative ways of knowing and being human:

> [T]he subject of epistemic freedom cannot make sense without addressing the key question of what it means to be human. This is the case because denial of being necessarily meant rejection of epistemic virtue.

Euromodernity was fundamentally an attack on human freedom even though it claimed advancement of human freedom. What the decolonization and deprovincialization struggles of today are grappling with is not just an epistemic issue cascading from the crisis of reason but rather an ontological question emerging at the center of dismemberment and dehumanization. Consequently, the crisis of Euromodernity has been the invention of and naturalization of paradigm of difference, which instantiated a politics of alterity predicated on the notion of the unbridgeable distance between those who called themselves European and those who became variously named as tribes, indigenous, primitive and black people.

(ibid: 243–244)

Cultivating such dialogues and transforming political and spiritual movements is an aspect of our responsibility. Along with epistemic freedom, we also need to cultivate epistemic responsibility. Responsibility here implicates both the colonizers and the colonized, Eurocentric as well as onto-decolonial thinkers and self and others. Ndlovu-Gatsheni's project of epistemic freedom has an incipient project of epistemic responsibility which needs further elaboration and cultivation. Epistemic responsibility is part of multi-dimensional movements of critique which goes beyond the discourses of rights and justice including cognitive justice. Responsibility refers not only to a static state but also to a dynamic process of both action and meditation, what can be called meditative verbs of co-realization and pluralization. Epistemic responsibility is part of processes of meditative verbs of pluralization which includes meditating with de-colonial possibilities even in colonial and Eurocentric epistemologies. Practice of de-colonial love is not only our goal but also our path which also calls for continued efforts to go beyond dualisms of self and other. Responsibilization embodies such movements of overcoming dualism and realizing conditions of mutual learning and blossoming with and beyond exiting epistemic bondage and colonial violence. Epistemic responsibility, as part of this broader movement of responsibilization, is simultaneously epistemic, ontological, and ecological;[24] ethical and aesthetic; political and spiritual.

Global responsibility involves realizing epistemic freedom not only in Africa but also all across the world. Along with epistemic freedom, we also need to cultivate pathways of epistemic responsibility. Epistemic responsibility is also confronted with the challenges of cultivating ways of knowing with and beyond anthropocentrism embodying both multi-species co-being and multi-species knowing. We can cultivate new ways of knowing bringing both human and non-human ways together. This opens new chapters in what Habermas calls knowledge and human interest and moving towards multiple ways of knowledge, freedom, responsibility, and planetary realizations (Giri 2013).

Notes

1 As Richa Gautam and Anju Singh (2010) write: "The CSR movement was an early response to an article published in 1970 by Friedman stating that 'social responsibility of business is to increase its profits.'"

2 Being in this conference, I had composed the following two poems as a way of reflecting on the calling of responsibility and the Anthropocene.

Poem One

Dark Hope

You say there is no hope
What is hope?
What is its color?
Is it only white?
Red, purple or blue?
Is it also not black?
Is it also not dark?
Is there hope in darkness?
Does hope fall from the sky?
Do we have to keep our mouth
Ready to suck the nipples of hope
Or do we have to pray and work
Love and meditate
With light and darkness
Relationship is the soil of Hope
Soil and Soul dancing together
Kissing each other
With hope towards hope.

> [On the train from Abadia
> Monteserrat, January 18,
> 2013, 5:30 PM]

Poem Two

Island Perspectives

What would you bring to the island
To island perspectives
One brought a cat
Another a rock
Yet another a feather
Oh traveler
Courageous painter of questions
What would you
Bring to the island?
To island perspectives
I would bring a boat
An island is not just an island
An island is also a relation
A field of communication
In the ocean to move in between and across
The boat is our companion
It is not just a medium
Nor just a passage
The boat is our mother

To the island perspectives
I would also bring my mother
The mother within
The mother without
The other as mother
The boat as mother
The ocean as mother
The world as mother
The cosmos as mother
Our islands becoming
Motherly passages of communication
Island perspectives
Impregnating intersubjectivity and transsubjectivity
Dancing together in a new yoga of transformation.
                              [Montesserat, January 18, 2013]

   See Giri 2019b.
3 Chakraborty quotes Amartya Sen:

> Consider our responsibilities toward the species that are threatened with destruction. We may attach importance to the preservation of these species not merely because the presence of these species in the world may sometimes enhance our own living standards. . . . This is where Gautama Buddha's argument, presented in Sutta Nipata, becomes directly and immediately relevant. He argued that the mother has responsibility toward her child not merely because she had generated her, but also because she can do many things for the child that the child cannot itself do. . . . In the environmental context it can be argued that since we are enormously more powerful than other species, . . . [this can be a ground for our] taking fiduciary responsibility for other creatures on whose lives we can have a powerful influence.
>
> (2021: 147–148)

4 As Chakraborty writes: "The climate crisis has brought the planet into the view, but we don't have a planetary form of governance" (Chakraborty 2021: 215).
5 Here what Strydom writes helps us:

> The acquisition of a cognitively fluid mind was a precondition for meeting the climate change challenge of 10,000 years ago. It was central to the solution which consisted of a momentous transition to a new sedentary social form of life, yet what cannot be overlooked is that this form of life soon in the course of its emergence practically took on a number of related negative features. Over the subsequent millennia, this sedentary form of life animated by the cognitively fluid mind and realized through a complex of practices proved to provide the basic ingredients for the emergence of modernity, including the contemporary global version of a reflexive modernity. The latter is characterized by a series of advanced properties which could be traced back to the cognitively fluid mind, yet these nevertheless suffer from negative features comparable to those associated with the original sedentary form. On investigation, it becomes apparent that while the mind has become cognitively flexible so that everything could be related to everything else, this capacity to open up and to facilitate connections in multiple directions has not been sufficiently translated into the actual organization of social life.
>
> (2015: 13)

6 In the words of Piet Strydom:

> Explanatory critique . . . requires the backing of an abduction-led norma-
> tive reconstruction which, in the first instance, allows the identification of
> a problem situation and the diagnosis of the social pathology at its core. Its
> central concern, however, is the relation of determination manifested in the
> historically sedimented second nature in the form, say, of an unjustifiable
> or badly justified social practice, institution, system or cultural model. To
> recall, persistent problematic features of modern society which undoubt-
> edly contribute disproportionately to climate change . . . such as the rai-
> son d'e'tat-driven nation state, the exploitative capitalist economy, the
> over-complex global financial system, the fat-cat global corporation, biased
> law, a technologically and capitalistically deformed science, and reified and
> distorted orientation and justification complexes for practices. More spe-
> cifically, however, the focus of explanatory critique is trained on the causal
> mechanism or mechanisms behind these problematic features.
>
> (2015: 16–17)

7 This and the following paragraph build on a webinar discussion on University
Social Responsibility organized by UK India Business Council in partnership with
University of East Anglia, U.K., and University of Hyderabad on April 16, 2020.

8 The noted political theorist Fred Dallmayr (personal communication) makes this
distinction between civic pandemic and viral pandemic. Continuing his thoughts
presented in the epigraph of this essay, Dallmayr tells us how political leaders
manipulate the pandemic in countries like USA: "Medical experts or experts in
medical science are often overruled or pushed aside by political power-seekers.
In this way, the two epidemics become one big threat to humankind" (personal
communication).

9 As Memembe (2020) writes:

> Before this virus, humanity was already threatened with suffocation. If war
> there must be, it cannot so much be against a specific virus as against every-
> thing that condemns the majority of humankind to a premature cessation
> of breathing, everything that fundamentally attacks the respiratory tract,
> everything that, in the long reign of capitalism, has constrained entire seg-
> ments of the world population, entire races, to a difficult, panting breath
> and life of oppression. To come through this constriction would mean that
> we conceive of breathing beyond its purely biological aspect, and instead as
> that which we hold in-common, that which, by definition, eludes all calcula-
> tion. By which I mean, the universal right to breath.

10 We can here relate the rise of Corona virus from destruction of wild habitats
to the burning of Khandava forest in Indian epic Mahabharata. Lord Krishna
and his companion heroic Arjuna burnt the Khandava forest to propitiate the
desire of Agni, the god of fire. All animals, bird, and inhabitants of the forest
were killed including a baby snake from her mother's womb chased and killed
by the great Krishna Arjuna duo. Tatkshaka, the king of snakes, escaped and
he built a magical palace with the help of Maya for the Pandavas seeing which
King Duryodhana became envious. He then defeated the Pandavas in the game
of dice and then humiliated Draupadi. Draupadi tried to seek revenge which
was one of the reasons of the Mahabharata war. According to Irawati Karve, it
is the burning of the Khandava forest which is the prime reason for subsequent
violence in Mahabharata including violence against Draupadi helping us to realize the significance of works by scholars such as Vandana Shiva that violence
against women and Nature go together (Karve 1991; Shiva 1989). But while

Draupadi took revenge against her humiliation, she did not take to task her husband Arjuna and friend Krishna for the degradation of Nature, for the burning of Khandava forest. To complete the story of revenge, Takshaka finally bit her grandson King Parikshit. But before this the King wanted to listen to words of wisdom from the sage and writer Vyasadeva who wrote Srimad Bhagavatam. In a similar way, we can realize the arrival of Corona virus from our destruction of forest and wild habitats and we need to compose and listen to a new Bhagavatam which is a story of human, nature, and divine and act wisely by reversing and transforming our ways of destruction. President Jair Bolsonaro who was infected with Corona virus has been involved with the burning of the Amazon rain forest. Now the virus has caught on him. And it is out to catch all of us unless we stop our thoughts and actions of burning our forests and other beings and weave our own paths and stories of human, nature, and divine, a new Bhagavatam, a new Corona, and *Karuna* (compassion) Bhagavatam.

11 In the famous words of Habermas (2020): "Never before has so much been known about what we do not know."

12 Here what Veena Das (2020a) writes deserves our careful consideration:

> How could the government not see and only realize belatedly that the policy of lockdown was directly contradicted by the offer of free buses to ply migrants across the border? And not only that the crowds gathered there would pose immediate risks of infection among themselves, but also that as these migrants spread out in villages, it would become impossible to trace contacts? Why did the higher-ups in the police administration neither think that policemen on patrol needed masks and gloves, nor that one stern order to the effect that anyone found using *lathis* (long wooden sticks) to beat up people would be suspended, or a one-day tour of affected areas by senior police officers to rein in the lower-level policemen might have constrained them from using their sticks so freely?

13 Here what Upendra Baxi (2020) writes:

> Epistemic humility is then a first virtue of the tasks and labours of governance, development, rights and justice. What matters is not the willingness to strike accompanied by the fear of being wounded, but the willingness and ability on all sides to learn from manifest and latent mistakes in law, policy, and governance accompanied by rapid course-correction. Epistemic humility requires us all to learn from and listen to each other, and not take recourse to antagonistic politics that divides people into the friend and the enemy. Second, not just the social activists but the bureaucracy, security, and political elites need always to accept the accumulated constitutional wisdom of the judiciary. A subservient judiciary and a sycophantic legal profession (and media) are no friends of a democratic order. If the Supreme Court is to remain a "last resort of the bewildered and the oppressed" (in the immortal words of Justice P K Goswami), adjudicatory co-governance of the nation (the wisdom of the people speaking through courts, or demosprudence) is as essential as is legisprudence (the wisdom of the legislators) and jurisprudence (the wisdom of the jurists). Pooling these wisdoms is the need of the hour in combating COVID-19. Third, this pandemic renders political oneupmanship or hyperpartisanship entirely superfluous and downright dangerous. Epistemic humility requires us all to learn from and listen to each other, and not take recourse to antagonistic politics that divides people into the friend and the enemy. If nothing else, COVID-19 should teach us to care for each other and co-learn from fellow-citizens, rather than use the arsenal of law to silence or punish free and responsible acts of speech and expression, howsoever

inconvenient or irritating these may prove to those who hold power. To accuse each other of defamation, hurting religious or cultural sensitivities, sedition or even treason is an undemocratic vice not to be celebrated.

14 Jahanbegloo (2020: 62) here presents the thoughts of Jean Keane which is important in our engagement with responsibility in this book:

> Responsibility is the center of gravity of the self. . . . [I]t requires the existence of a person who is responsible and someone or something for whom or for which that person is responsible. Responsibility orientates the self—endows that self with the power to confer meaning on its relations with the wider world. . . . That means that the struggle to establish and cultivate responsibility is a life-and-death struggle to survive as a human being. Life is constantly threatened by nothingness . . . the power of invididuals . . . to fend off nothingness, depends upon their . . . capacity for responsibility.
> (Keane 2000: 289–290)

15 Here what critical economist Irene van Staveren writes is also helpful:

> I think if we look at the world economy as a whole, we see that most world trade is dominated by multi-national companies, by intercompany trade. That is because huge specialization in low cost production along very long value chains. The corona crisis has shown us the vulnerability of such massive specialization: the global North is too dependent on just a few value chains for key products, while the economies in the global South are dependent for their employment on those same value chains. This calls for a systematic change to the world economy towards shorter supply chains.
> (van Staveren 2020: 19)

16 Pope Francis speaks about being apostolic in our listening to each other. Inspired by this I have written this poem which can be read in our journey of cultivating a new poetics of apostolic and listening leadership:

Apostle of Ears

Ananta Kumar Giri

Apostasy
Apostle of Fear
Where are Apostles of Ears?
Marching in the Name of Kingdom of God
Ram Rajya—Kingdom of Rama
Banishing Sita and Killing Shambuka
On the Way
Sacrificing Innocents as Lambs
Where are your tears
Where are your Ears?

[I dedicate this to Pope Francis and other sadhakas and sadhikas of listening in this anxiety-ridden and apathetic world of ours. Bangalore, September 4, 2017. This is in the author's book of poems, *Alphabets of Creation: Taking God to Bed*. Giri 2022a]

17 Here as Butler (2020) challenges us: "We would need to develop political practices to make decisions about how to live together less violently."

18 Here what Veena Das (2020a) writes with a personal touch is an inspiring example of ordinary ethics:

> I am a realist. I know that I belong to one of the "vulnerable" groups, and indeed, in the triage of hospital beds or ventilators, I would rather that a

younger person with more life to live gets priority over me. Yet I do what I can to survive.

19 In the context of Coronavirus, what Ramin Jahanbegloo (2020: 1) writes about living with an art of listening and singing like birds and living wisely is helpful:

> We all live the same life but we each live it differently. That is what makes life interesting. Starting the day at dawn is an art that the birds have perfected. They sing to us innocently and without hesitation, not knowing that we human beings have lost the art of living. If we human beings keep our faith in life, if we believe in living with equal faith, we will know how to live like the rest of the natural world. It is by a mathematical point only that we are alive today, but mathematicians don't know how to listen to the birds singing. We may not be able to give meanings to our lives with calculus or trigonometry, but we can certainly maintain ourselves on this earth by living according to the dictates of wisdom. To be wise is not merely to follow the path of reason. Many people are capable of common sense, without being necessarily wise. To be wise is to see the cruelty of fate, but also to be able to surpass it.

One implication of the aforementioned thought of Jahanbegloo is that we have to learn to live with Coronavirus wisely not accepting it as a cruel fate. We have to learn how to live with this, making our fate into a destiny—paths of mutual destination with love, care, wisdom, and responsibility. The same calling is also there with pathways of death. We also have to learn how to embrace death with wisdom while doing all that we must do to nourish and protect life.

20 Veena Das (2003) here challenges us to overcome the dualism between sociology and common sense and how sociology can learn together with common sense for deeper understanding of self and society.

21 Following Horton's epigraph about this, here what he writes is also challenging:

> At press conference after press conference, government ministers and their medical and scientific advisors described the deaths of their neighbors as "unfortunate." But these were not unfortunate deaths. They were not unlucky, inappropriate or even regrettable. Every death was evidence of systematic government misconduct—reckless acts of omission that constituted breaches in the duties of public office.

22 Living with fragility during our Corona times also is an invitation for us to cultivate our immortality with and beyond our mortality which becomes our sources of hopes. Here we can draw inspiration from Sri Aurobindo's epic *Savitri*:

> Born of its amour with eternity
> Our spirits break free from their environment.
> The future brings its face of miracle near,
> Its godhead looks at us with present eyes;
> Acts deemed impossible grow natural;
> We feel the hero's immortality;
> The courage and the strength death cannot touch
> Awake in limbs that are mortal, hearts that fail;
> We move by the rapid impulse of a will
> That scorns the tardy trudge of mortal time.
> These promptings come not from an alien Sphere:
> Ourselves are citizens of that mother State.
> (Sri Aurobindo 1993: 262)

23 Escobar (2021: 3) presents us the discussion about Terricide or killing of Mother Earth taking place in Latin America and in the web:

We define terricide as the killing of tangible ecosystems, the spiritual ecosystem, and that of the pueblos (peoples) and of all forms of life. Confronted with the terricide, we declare ourselves to be in permanent struggle, resistance, and re-existence against this system. . . . We summon all peoples to build a new civilizational matrix that embraces Buen Vivir [good living, collective wellbeing] as a right. Buen Vivir implies the retrieval of harmony and reciprocity among peoples and with the Earth. Summoned by the memory of our ancestors and the lands and landscapes that inhabit us, we have agreed on the creation of the Movement of Pueblos against Terricide.

24 We can remember here the works of Gregory Bateson (1972).

5 The Calling of Global Responsibility[1] and the Challenges of Planetary Co-Realizations

The plight described is the ultimate consequence of substituting competition and rivalry—the mode of being derived from belief in the greed-guided enrichment of the few as the royal road to the well-being of all—for the human, all-too-human longing for cohabitation resting on friendly cooperation, mutuality, sharing, reciprocated trust, recognition and respect.

Zygmunt Bauman (2013), *Does the Richness of the Few Benefit Us All?*, p. 90

The concept of a just world is an aggregate profile of the world in political, economic, ecological, social and cultural terms where conditions of justice prevail at global, regional, national and local levels. Besides being comprehensive in scope, it clearly links up various spatial levels: from the grassroots to national territories to continental and global levels. In other words, in a just word, every group, every region, every culture would live in conditions of justice. On the other hand, global justice movements have been issue-oriented and in many ways fragmented.

Yet the global trend of people's rights movements demanding justice and equality and freedom has forced elites of all countries to accommodate demands for human rights and environmental justice. What is, however, important to note is that the discourse of global justice is vulnerable to be incorporated within the existing world, political and economic order. On the other hand, discourse on a just world clearly affirms the need for transforming the unequal and unjust world into a just world. It is no doubt going to be a long historical process, but what is important is clarity of vision.

—Manoranjan Mohanty et al. (2015), "Introduction," *Building a Just World*

We know the pain points. Everyone knows we must end extreme poverty for billions. Everyone knows we must fix the inequality crisis. Everyone knows we need an energy revolution. Everyone knows our industrial diets are killing us, and the way we farm food is ripping through nature, driving a sixth mass extinction of species. We know human populations cannot increase endlessly. And we know our material footprint cannot expand infinitely on our finite, blue and green Earth.

Can "we"—meaning all people and peoples—come together to navigate this century? Can we take a collective leap in human development with

DOI: 10.4324/9780429347481-5

courage and conviction? Can we overcome divisions, neocolonial and financial exploitation, historic inequalities, and deep distrust among nations to deal with the long-term emergency? Can we achieve systemic transformation in decades, not centuries?

Our goal with Earth for All is to show you that this is indeed fully possible. And that it won't cost the Earth. Rather, it is an investment in our future [. . .]

—Sandrine Dixon-Decleve et al. (2022), *Earth for All: A Survival Guide for Humanity*, pp. 1–2

In this study, we have looked into varieties of issues related to global responsibility such as global justice, cross-cultural and inter-religious dialogues, climate change, Anthropocene, corporate social and spiritual responsibility, COVID-19, and epistemic responsibility. It shows how we need to rethink existing theories and practices in order to address contemporary global challenges.

As part of this rethinking, Heikki Patomaki, a noted thinker of our times, proposes a different financial and economic architecture for the world rather than continuing the neo-liberal economic and social model. Reflecting upon the European Economic Crises that has engulfed countries like Greece, Spain, and Italy which he calls *The Great Eurozone Disaster*, Patomaki (2013) calls for putting in place a global Keynesianism. He pleads for more global democracy and a new global imaginary where we think and act simultaneously locally, regionally, nationally, and in the world with a new spirit of cosmopolitanism. For Patomaki,

> Global Keynesianism is an approach that frames questions of public economic policy and politics more generally on the world economic scale. Global Keynesianism aims to regulate global interdependencies in such a way as to produce stable and high levels of growth, employment, and welfare for everyone and everywhere, simultaneously. Global Keynesianism is an ecologically responsible doctrine: governing interdependence could not be otherwise sustainable. The main themes of global Keynesianism are public administration, democratic politics, mixed economy, global taxation, global redistribution of wealth, global aggregate demand, joint management of investments and financing, ecological sustainability, and the many levels and contexts of governance and their interconnections.
>
> (Patomaki 2013: 175)

The rising economic inequality in the world is a source of major turbulence which is creating crises as suggested in Thomas Piketty's work *Capital in 21st Century* (Piketty 2013). Global Keynesianism strives to address this. It works on redistribution of wealth, facilitative regulation, creation of employment through varieties of means, increase of real wages everywhere, and support for trade union movements and a new civilization and for a planetary economic policy. In his recent work on critique of contemporary global political economy, Patomaki (2017) links the proposal of Global Keynesianism with a new kind of reflexivity. We need to cultivate new visions, movements, and practices of what Patomaki calls "ethical and political" learning and "heteroreflexivity" to which we can also add aesthetic and spiritual co-learning. We can realize the significance of different movements of global justice, dialogues, and responsibility discussed in this book as contributing to ethical, political, and spiritual learning. It also calls for experimental creativity in building of creative self and institutions in social, economic, and political spheres as part of a new civilizational realization for not only humanity but also life going beyond Eurocentrism, ethnocentrism, anthropocentrism, and the dualism between civilization and barbarism (see Unger 2004).

Global responsibility is thus linked to a new civilizational experimental creativity at self, societal, national, world, and planetary levels. In this context, Patomaki tells us that The Brandt Commission on Sustainable Development "developed the idea of a world civilization for the new millennium, and proposed a new international economic system" (Patomaki 2013: 175–176). The civilizational dimension of the present crises and movements for renewal are also highlighted in the work of Jeffrey Sachs (2012) who titles his reflections on the current crises as *The Price of Civilization*. The present crisis is part of a crisis of civilization, of Western Civilization of modernity in particular (Eisenstadt 2009 had coined the term civilization of modernity and Gandhi (1909), had challenged us to realize some of the foundational ills of this civilization of modernity). But to sustain civilization Sachs says that we need appropriate taxation. The rise of the political Right in the Euro-American world has cut down taxes which have led to production of bare life and barbarism in terms of social suffering and dismantling of services. An appropriate taxation is not confined only at the national level and it should be thought of at the global level too. In fact, the Brandt Commission Report had argued for global taxation, the revenue for which would be used in efforts to eradicate poverty and to promote economic development of the global South (ibid: 176). Thomas Piketty (2013) also emphasizes upon appropriate taxation not only at the national level but also at the global level so that humanity does not descend into barbarism of unsustainable inequality and consequent violence and destruction of life.[1] For this, Piketty calls for the rise of a new social state based upon impulse of political economy which is interested in redistribution of income and capital and strives for realization of wide-spread social well-being. Different global justice movements described in this book such as ATTAC and Ekta Parishad share this.

In his recent work, *Capital and Ideology* (2020a), Piketty tells us that ideological inequality is not just dependent upon economic system but is produced by ideological structures which justifies inequality and if we are able to create new ideological movements for equality then we can transform contemporary conditions of political, economic, and social inequality and create conditions and movements of equality. Here Piketty presents two possible pathways towards future: social federalism and participatory socialism. For Piketty,

> Social federalism is a view that if you want to keep globalization going and you want to avoid this retreat to nationalism and the frontier of the nation-state that we see in a number of countries, you need to organize globalization in a more social way. If you want to have international treaties between European countries and Canada and the U.S. and Latin America and Africa, these treaties cannot simply be about free trade and free capital flow. They need to set some target in terms of equitable growth and equitable development. "But this does not strike at the structural bases of inequality." Participatory socialism, on the other hand, strives to address some of these.

For Piketty (2020b), "Participatory socialism is the general objective of more "access" to education. Educational justice is very important in terms of access to higher education. Today there's a lot of hyper criticism, not only in the U.S., but also in France and in Europe, that we don't set quantifiable and verifiable targets in terms of how children [from] lower [income] groups [gain] access to higher education, what kind of funding [they] have for higher education. The other big dimension is circulation of property, so I talk about 'inheritance for all'" Piketty here proposes a "progressive tax on wealth in order to finance [a] capital transfer to every young adult at the age of 25. This transfer is in effect, 120,000 euros [about $134,000] per person, [which is] about the level of medium wealth today in France or in the U.S. That will very much transform the ability of children from poor families or middle-class families to create their own firms."

But what is striking is that, while talking about taxation and the need for a new social state, Piketty does not refer to the work of any social movements or tax justice movements such as ATTAC which despite his call for new forms of organization[2] makes his perspective a bit statist though Piketty himself is a member of ATTAC. We need to link the proposal of creation of a new social state with dynamics of movements where movements are not only socio-political and socio-economic but also socio-spiritual combining the impulse of political economy with moral economy, moral sociology, and spiritual ecology (Giri 2021b). As has been discussed in this study, an appropriate global taxation is also part of many global justice movements such as ATTAC. But the pathways towards a new civilization which Sachs, Piketty, Mohanty, and Patomaki hint at also call for cultivation of such values as empathy and mindfulness. Jeremy Rifkin (2010) in his work *Empathic Civilization* draws our attention to it as does the economist Sachs who challenges us to realize the need for examining the life we are leading and cultivation of a mode of mindful living in economics and society. As Sachs (2012: 9) writes:

"The Unexamined Life is not worth living," said Socrates. We might equally say that the unexamined economy is not capable of securing our well-being. Our greatest national illusion is that a healthy society can be organized around the single-minded pursuit of wealth. The ferocity of the quest for wealth throughout society has left Americans exhausted and deprived of the benefits of social trust, honesty and compassion.

We can escape our economic illusions by creating a *mindful society*, one that promotes the personal virtues of self-awareness and moderation, with the civic virtues of compassion for others and the ability to co-operate across the borders of class, race, religion, and geography.

Sachs' pointer to mindful economy reminds us of the work of sociologist Robert Bellah who challenges us to create a mindful society (Bellah et al. 1991). For Bellah, a mindful society is one which is a society of attentiveness rather than distraction. This involves critiques and creativity of political economy, moral economy, and moral sociology. This is not achieved by only focusing on issues restructuring state and economy but by work on self-transformation which is suggested in Sachs and Rifkin. Gandhi and many other seekers also point to the need for cultivating an attitude of restraint and non-possession which would help us to create a new civilization required for our present-day world. Global responsibility calls for habits of mindfulness at local, national, global, and planetary levels.

Thus, Patomaki's proposal of Global Keynesianism and Piketty's proposal of a new social state, social federalism, and participatory socialism need to be part of related multiple processes of transformations such as transformation of our economic thinking and practice along lines of restraint and non-possession (what Gandhi calls *aparigraha*) and reduction and transformation of consumption and production as raised by the de-growth movement in Europe. The proponents and activists of the movement urge us to realize that "overconsumption lies at the heart of long-term environmental issues and social inequalities."[3] It also calls for locally sustainable production and consumption which is also reflected in the growing localization movement. For example, in Bangalore, Bhoomi College had organized an international conference on localization in 2013 in which many of the participants stressed the need for localizing production and consumption and be in engaged with creative works on self, culture, and society. The participants included authors such as Helena Norberg-Hodge (1991) whose book *Ancient Futures* has been a source of inspiration to many to think beyond the present model of modernization and development. It also included P.V. Rajagopal who is also striving for creative localization and alternative globalization carrying the seeds of what Bruno Latour calls localization plus and globalization plus. Manish Jain has been nurturing a different educational path called Swaraj University in Udaipur, India, which is a radical experiment in education where learners who are called *khoji* (seekers) come from different backgrounds and learn with local people by walking and travelling around with them. Localization movements need also to be part of the restructuring of global economy and politics which also involves interrogation and transformation of existing models and organizations of market and state. The proposal of global governance and global economic reform needs to be part of transformation of state and market through multi-dimensional processes of self, social, cultural, economic, political, and spiritual transformations. Global Keynesianism and proposed architecture of global governance thus need to be part of transformation of the very logic of state, market, economy, exchange, and power. It thus needs to be accompanied by what many thinkers and movements have been striving for: transformation of state power into multiple powers of self, societies, and communities as well as transformation of the very logic of power as one of creating new realities and possibilities together rather than continuing a project of domination over others. It seems the present discourse of global Keynesianism and global governance has not sufficiently attended to these issues of self-critique and mutual generativity. Global governance needs to practice varieties of modes of self-governance and mutual governance at the level of self, societies, and globality bringing Gandhi and Foucault together in transformative ways. This involves a new hermeneutics of self and a new modality of government at the root of which lies self-government (Foucault 2005). The discourse of global governance thus needs to bring the vision and practice of Swaraj and needs to talk about what Manoranjan Mohanty calls *Jag Swaraj* or Global *Swaraj*. It also involves cultivation of what Ashish Kothari calls Eco Swaraj (see Kothari et al. 2019). This involves a fundamental transformation of state practice and statitst ways of thinking, as it involves ethical, aesthetic, and spiritual critique and renewal of self, state, society, and polity.[4] In this context, realization of global responsibility is intimately linked to going beyond one-dimensional concepts of global governance, economy, politics, justice and universality, and creating a pluriverse (Escobar 2008, 2018, 2020).

Realization of global responsibility is thus part of manifold processes of social, political, and spiritual transformations. These include simultaneous movements in justice and dialogues, ethics and aesthetics, politics and spirituality. These processes of transformations help us realize each other and become part of manifold processes of co-realizations facilitating birth of rainbows of planetary co-realizations going with and beyond entrenched closures of many kinds such as state-centred rationality, Eurocentrism, ethnocentrism, and religious fundamentalism. In this study, we have explored various such initiatives and movements in global responsibility, which help us realize webs and colours of planetary realizations and co-realizations. The visions and practices of these movements hold signs of hope that we can build a different world of beauty, dignity, and dialogues for all—indeed An Earth for All (cf. Dixon-Decleve et al. 2022; Francis 2015).

Notes

1 Piketty (2013: 471) writes: "But a truly global tax on capital is no doubt a utopian ideal. Short of that, a regional or continental tax might be tried, in particular in Europe, starting with countries willing to accept such a tax."
2 Piketty urges us to realize that "new forms of organization and ownership remain to be invented" (2013: 403). Piketty (ibid: 482) here presents us a glimpse of the future:

> [N]ew decentralized and participatory forms of organization will be developed, along with innovative types of governance, so that a much larger "public sector" than exists today can be operated efficiently. The very notion of "public sector" is in any case reductive: the fact that a service is publicly financed does not mean that it is produced by people directly employed by the state or other public entities. In education and health, services are provided by many kinds of organizations, including foundations and associations, which are in fact intermediate forms between the state and private enterprises.

3 From the internet article on it.
4 Such a challenge of linked process of transformation of state which needs to accompany projects of global reforms such as global governance or global Keynesianism is suggested in the following poem by the author:

> State
> Political and economic
> Caste, class and gender
> Movements and self
> Transforming machineries of violence
> Ethical Critique and a New Aesthetics of Commons
> Collective Action and Collaborative Imagination
> Transforming State from Within and Across
> State and Spirit
> Spirit and State
> Dancing in an Open Way
> With Hegel, Kierkegaard, Gandhi and Sri Aurobindo
> Going Beyond the Divide Between History and Pre-History
> Towards a New Unfoldment of Potential
> State Becoming a Non-State
> A Non-State State
> As Self Becomes a No-Self.
> Democracy as Political and Spiritual.
> Toward a New Tapasya of Co-Emptying
> And evolution of consciousness
> A Festival of Co-Realizations.
>
> (Giri 2022a)

The Calling of Global Responsibility

A Roundtable of Epilogues and Conversations

Global Responsibility and Global Humanization

An Epilogue

Fred Dallmayr, University of Notre Dame, USA

My friend Ananta Kumar Giri has asked me to write a brief epilogue or afterword to his book, *The Calling of Global Responsibility*. I am happy to do so. I have known Ananta personally for many years, during the time when I was travelling extensively throughout India, Asia, and other parts of the world. Now during the global pandemic, travelling has become nearly impossible. However, his book can serve nearly as a substitute for far-flung travels. For, what is intimated by his text is that, despite the pandemic, the big world is still there and, more importantly, that a new world is beginning to emerge behind the screen of everyday habits and routines.

The new world which is emerging—and Giri's book allows us to peak into the process—is indeed an un-habitual world, one not firmly tied down to traditional structures or conventions. The ongoing process is still fluid or in flux and does not allow for neat definitions. However, broadly speaking, one might say that the "old world" which is receding was structured "top-down," that is, anchored in traditional states and their traditional elites. By contrast, the "new world" whose birth pangs we are witnessing is built "from the ground up," that is, anchored in loosely structured social movements and grassroots gatherings whose agendas are often inchoate or evolving. Whereas the older structures were firmly tied to existing boundaries or borders, the newer gatherings are perhaps not "borderless" but uncomfortable with neat dividing lines.

In many ways, Giri's book is an exercise in multiple border-crossings. One border is the line dividing national cultures and languages. While traditional academic books pay tribute to one or a few cultures and languages, Giri's text is an introduction to world literature on the topic of global responsibility and public justice. The list of scholarly names invoked in the text is dizzying and mind-blowing. Thus, one encounters such well-known names as Apel, Ricoeur, Nussbaum, Pogge, and Derrida, but also such less familiar names (in the West), such as R. Sundara Rajan, Shiv Vishvanathan, P.V. Rajagopal, and J.N. Mohanty. The point of this global nomenclature is not to impress readers with super-expertise but to promote a deeper goal: that of global humanization. As Giri writes (p. 6):

> Globalization as humanization strives for a fuller realization and for integral development (spiritual, [political. Economic and social) of self and

DOI: 10.4324/9780429347481-7

society. . . . Planetary realizations refer to a new cosmopolitan awareness that all of us are children of Mother Earth and calls for transformation of existing boundaries of rationality, nation-state and anthropocentrism.

On a more concrete and practical level, Giri's text analyses a series of new grassroots movements which have sprung up all over the world roughly during the past 50 years. Readers not familiar with these developments will find these discussions particularly valuable. Many of the movements can be understood as protests against "neo-liberal" capitalist forms of global domination or as strivings for realization of greater social justice. For example, Giri introduces readers to the important work of ATTAC, a movement operating in many countries striving for greater tax justice and social solidarity. In India, we learn about the social movement of Ekta Parishad which draws much of its inspiration from Gandhi's work with tribals or Adivasis. On a broader scale, we find the activities of the "World Social Forum" which seeks to create for people a livable world in lieu of what Jean-Luc Nancy has called an "unworld." Special attention is devoted in the text to cross-cultural and inter-religious dialogues in Indonesia and their academic cultivation at the Percik research institute. Also discussed are more formal institutions like the UN Alliance of Civilizations and numerous similar organizations.

As I mentioned before, the chapters of the book do not offer a complete picture of the emerging "New World" (which would be quite impossible today), but they all point intriguingly in the new direction. As Giri states (p. 135), what we have so far are signs of "a fundamental transformation of state practices and statist ways of thinking as it involves ethical, aesthetic and spiritual critique and the renewal of self, society and polity." Needless to say, it is by no means clear or predictable whether the agents of the "new world" will be less susceptible to corruption, domination, and violence than the agents of the "old." This question cannot be resolved by scholars in their books. All depends on proper education, exhortation, and guidance by good example—which is the global responsibility of us all.

Responsibility and Global Governance

P. V. Rajagopal, Ekta Parishad, and Jai Jagat

I have gone through the work of Professor Ananta Kumar Giri on the subject of responsibility and global governance and I do greatly appreciate his prodigious efforts. Whenever I have met Dr. Giri and heard about his visits to various countries, attending different conferences and engaging people in discussions and reflections, I had no idea that he could so imaginatively use all these raw materials to analyse current social issues and provide some direction for those of us working in the social sector. I want to congratulate him for the insights he has provided in his book.

I have a deep affinity with Dr Giri's view of a world order that can transcend nationalism and not be restricted within artificial boundaries with the false notion of the permanency of its own statehood—nation state. I paraphrase Dr. Yuval Harare when he says that "nations and nationalism are all based on some kind of an imagination developed by clever people that are keen to keep large numbers of people (*Homo sapiens*) together." He goes on to show how people use imagination to develop religion as well as nation states, and in the cases, that states were ruled by kings they were deemed the representatives of God while in other cases, self-appointed authoritarian leaders gain their legitimacy through elections. In either case, they continue to rule a territory or geographical region in the name of people that are oftentimes victims of their whimsical behaviour and decisions.

In human history, people crossed large geographical areas in search of habitat and opportunities. Human beings have expanded their territories by fighting wars and using all means of force. Those with the greatest might declared themselves as the rulers, and gradually as the human civilization evolved, people learned to respect the idea of self-determination and becoming free from the colonial rule of the British, Portuguese, French, Spanish, and the Dutch. The colonial powers were compelled to leave the territories that they had occupied by force.

Though governance by local people was a good idea in all cases, the people have not been able to withstand ruling elites and their powerful lobbies. Dominant groups within the country start governing and controlling the weak and voiceless majority in the name of sovereignty. This is subterfuged in what is called "democratically elected government" so even though this

DOI: 10.4324/9780429347481-8

appears to be social progress having people's representatives replace outsiders or dictators, it remains a crude form of democracy. A large number of people are not even aware of the degree of manipulation that is going on in their name with negligible benefit.

Using this newly found idea of nationalism or the nation state, the rulers are able to whip up a feeling of superiority of their people, and the importance of protecting its glory and national identity. This patriotic sentiment can be used by the rulers of the state to fabricate many laws and to control and punish the people whom the state is supposed to serve. The state can also oppress those who may show disagreement with the ideas or ideology of the ruler or the ruling class. It is evident that in many countries, there are a large numbers of people who are in the prison just because they opposed some of the policies proposed by the state. It is difficult to assess how many will have the same level of good fortune that Mr Nelson Mandela had in returning to civilian life and having the opportunity to provide a political alternative to their people.

Personally I do not agree with the idea of the "survival of the fittest" though what is happening in today confirms this old theory. Those with money power and muscle power continue to hold onto political power in many parts of the world and I hope the day will come when we can introduce a new theory of "making everyone fit for survival" rather than "survival of the fittest."

When we look at the world situation of the day from this angle, Giri's argument for a planetary realization and global responsibility is relevant and forward-looking. I think a call for global responsibility is the most important thing that should happen now. The adverse effects of the COVID-19 pandemic have taught us what can happen if we don't have responsible global governance system in place. It is time to decentralize much of the governance responsibility to the communities at the local level and also to give up part of the responsibility to enable an effective global governance system based on consultation and collaboration. Issues, such as poverty, inequality, the climate crisis, warmongering, or even this pandemic, cannot be tackled by any one country. It demands a responsible global governance system to deal with the serious issues of wide-spread poverty, distress migration, and large-scale human rights violations, as these cannot be left to whimsical and dangerous leaders at the national levels. This is the time to look forward and develop forms of governance that give people hope for the future rather than being paralysed by lethargy. There are many examples of irresponsible governance: whether it is ignoring the impact of COVID-19, countries using the opportunity to expand their territories or the lockdown of millions of migrant labourers. From these it is clear that there is an absence of a global system that challenges national leaders to be accountable. The distribution of power between three levels of government may improve the global governance system and also provide more justice to citizens and to the planet.

Recently I was reading a book on human behaviour and values. One point that attracted my attention was reaffirming the "golden rule" as globally acceptable, which is "you never do anything to others that you don't want others to do to you." The aggressive species that we humans are, with a history of capturing, controlling, and destroying, ought to move into a new stage of their evolution to become enlightened human being with a deeper commitment to the idea of sharing and caring. Sharing may come much later but at least to begin with an idea that I will not do harm to others, as I don't appreciate others doing harm to me. . . . I strongly feel that this should be one value that should be acceptable to all people and I hope this can be guaranteed through a responsible global governance system.

The challenge that the humanity is facing today is the capacity to dialogue. Leaders at different levels are using the language of power, threat of conflict and war, to address their problems rather than deepening their knowledge and skills in the art of dialogue. They are used to a language of arrogance and violence. In this process they also stir up nationalistic sentiments among their followers and create hate for those whom they oppose. In such situations, there are movements towards beginning wars and frozen conflicts. In the case of India and Pakistan after achieving freedom 72 years ago, they are not able to sit around a table and sort out their differences. The leadership of both countries continue to believe that war is a better strategy than dialogue. Another example is between North and South Korea. Unfortunately we live in a world where many leaders and their followers tend to believe that dialogue is for the weak, and war is for the strong. This theory is the base for many political leaders to stay in power. Whether we like it or not, people in many countries are conditioned to believe that when leaders are speaking the language of "might is right" looking tough and arrogant, that this best represents people's interest. The language of friendship and dialogue is perceived as weak. In a world run by violent politics and economy, it can only be changed or redirected through a globally responsible governance system that is backed by an equally responsible and accountable national system. Ultimately this is the only way to achieve peace and prosperity at the local and global level.

I am happy that Giri has spent a lot of time looking at some important NGOs and people's organizations such as ATTAC, Ekta Parishad, and some others. In his study, Giri has also looked at large NGO conglomerates like the World Social Forum. He has gone into great detail with each one of these organizations through interviews and field visits. As far as I can surmise, he is making efforts to assess whether these organizations have the capacity to influence the global processes in today's world. In my opinion, it is very important to gauge the capacity of non-state actors in influencing and directing responsible global governance. In fact in countries like India, voluntary organizations have done some pioneering work and the same can be said for their influence on policy-making. The land gift movement led by Vinoba Bhave is a very good example in this connection.

While Giri was exploring this question academically, I should admit that I was exploring the same question in many different ways through various ground-level activities. The question that worries me is how to make the governance system pro-poor? How can very ordinary people make their so-called elected representatives accountable? How can non-violent people's movements address larger policy issues at the state and the national level?

In order to find answers to these questions, we have tried three different approaches depending on the time and place of our work. The first method generally used was dialogue as we believed that the person in front could be ignorant of the situation or that he/she is not aware of a particular law that is being violated. It is important to talk to her/him before jumping to conclusions about their intentions. In some cases, this process of dialogue really worked, and in other situations, it did not yield any result. So we went for struggle. This was the second approach, that is to build pressure on the system so that they would listen to the voices of the marginalized. The third area is called constructive work as we believe that the achievements of the struggle should be converted into actions that improve the life of people that were part of the struggle. Here I should admit in line with what Giri describes in his chapter, that many of us have always returned from struggles with some success but never enough to change the life conditions of the poor and marginalized in a substantial way.

What is important to notice is not only direct achievements from the state but also the kind of training process that goes into organizing the poor to stand up in front of a gigantic state to act in order to bring about change in their life rather than expecting some representatives at the top to change their life situation without their own efforts. Breaking this dependency culture created through years of representative democracy is important. It is also important for people to realize their own power and act together is very important if we want to make the government or the system accountable.

What is important is also to understand why and how Jai Jagat ("Victory to the World") campaign was initiated? I think this is where one can find some answers to Giri's search for a responsible global governance system. The immediate question that comes to the mind of the reader is why would an organization with limited reach and resources plunge into a global action called *Jai Jagat* . . .? The answer to this question is likely to lead us to the central issue for which Giri is searching. Without creating a positive global environment through such global movements, no progressive idea whether it be "a border-free world" or responsible global governance, is going to happen. Though we began as an organization addressing local issues, as time went on, we grew to realize that we needed to play both at the local and global levels, if any meaningful change were to be brought about during our lifetime. Analogous to the arguments of Giri, this effort has to build a global justice movement rather than anti-globalization movement. The Jai Jagat campaign has set a positive agenda where there is a greater chance for individuals and organizations to come together. Anti-movements in our experience are less sustainable; at best they have temporary media hype or visibility. Today we see that it is becoming more acceptable to pull down some statues and loot some shops as a way to express the deep anger of being marginalized. This is the time to realize that building a responsible global system will involve everyone at every level. What is important is for the human beings (or Harare's Homo Sapiens) to move ahead from where they are today in the evolutionary stage and stop fighting and killing each other for more power and control.

Mahatma Gandhi during his lifetime was able to bring about revolution through small actions. How could anyone see revolution in a spinning wheel or in hand-full of salt? People had difficulty in believing in these small actions, yet they did ultimately lead to big change. Imagination demands that in small things, ordinary people can participate. In our work, we were also looking for the simplest thing that everyone could do without any special efforts. It was through years of experience that we came to the conclusion that the only participatory method of struggle for the poor people was foot march. Thousands of people walking together, suffering together became a very powerful tool and these are the kind of tools that can be used by ordinary people of any nation to make their government accountable. Imagine millions of people marching with a demand for change. Only by experiencing it, can one feel the power of the poor and also the power of walking and suffering. I have experienced this revolutionary power of a long-march and I am dreaming of a global one that could finally shake the international system and change the way it is organized.

The recent development in India is a clear example of the power of long-march though it was done by the unorganized people in an unexpected manner. Immediately after the announcement for a lockdown, we witnessed millions of poor labourers walking back to their villages. Even this unorganized walking was like a slap on the face of a government that was trying to project itself to the world as being a very successful government. If walking is not a powerful action, imagine then the reason the global media was speaking only about the people on the road for many days and not about any other issues. This is why I think people's movements across the globe should expand their imagination and invent small and feasible actions for people to participate.

I have noticed that Giri refers to the work of Stéphan Hessel and the work of Indignados inspired by Hessel in his book. As he was a friend of mine and I used to spend time with him discussing small methods that have the potential of bringing about non-violent change. No surprise his small book that was addressed to the youth created a kind of a revolution among the youth of France.

Non-violence is another area of importance in today's discussion. In a world where violence is a business and those who sell weapons of mass destruction are the richest people, it is clear that non-violent actors are up against a huge and violent system. Today's economic and development paradigm needs to be questioned and challenged by using non-violent methods. I am aware that civil society and social movements are divided on this issue. There is greater agreement that there needs to be transparency and accountability. For many, gender equality is important, but on the issue of non-violence, there is some kind of an ambiguity. This is the reason why it becomes difficult to pull everyone together for a global movement around non-violence. The forces that believe in violence as a method for action has also succeeded in educating and inspiring leaders of some social movements that their methods are superior. Unfortunately, as Gandhi said, "an eye for an eye leaves the whole world blind." Enlightened social movements that wish to sustain will definitely realize the importance of not shedding more blood and not spreading more hate and bringing people together around a purposeful sense of peace.

I want to congratulate Ananta Kumar Giri for his sincere and serious efforts in bringing some very interesting issues together for reflection. I am sure that this book will bring about greater understanding for a responsible and enlightened local to global governance.

Environmental Justice
Dialogues Towards Global Well-Being

Felix Padel, University of Sussex, UK

This book is a courageous survey of many crucial world arenas and philosophies of dialogue aimed at taking responsibility for creating a fairer world. The biggest challenge exists in our face every day, witnessing injustice on multiple fronts: how to take action to effect actual transformation of the structural violence? To paraphrase the words of Bernard Cassen (of ATTAC, quoted earlier): How to transform economic critique and grassroots movements—let alone ethical principles—into political action and real change?

If we examine, impartially, how we human beings have evolved, we have developed very far in technology, and in complexity of thought and social organization, but we have also developed hugely in war and cruelty, manifesting in seemingly insoluble conflicts in countless regions. Where we have *not* developed—despite some notable yet fragile breakthroughs—is in making peace in the face of war, and developing democratic power structures that ensure responsibility towards the ecosystems that future life depends on.

To progress further, surely we can agree that we need to evolve our sense of responsibility towards each other, and towards our natural environment that we inherit—ecosystems of great beauty and fecundity that our rapid industrialization and extractivism have seriously damaged and *de-developed*, threatening even further destruction in the imminent future.

The need here is for justice, that we can rely on impartially, without the inequality that comes from lawyers' fees and corruption among upholders of the law, such as police and security forces. The murder of George Floyd in America in May 2020 has brought into focus issues of racism and unjust policing evident in too many countries.

The cruel treatment of Julian Assange in the UK justice system, and of whistleblowers and journalists in many countries, highlights the huge obstacles that still block communicating truth to power in a way that can bring about real, structural change. Independent journalism is under attack in too many countries; as are environmental defenders, even though these are at the forefront of efforts to bequeath a healthier world to our grandchildren (see Lakhani 2020). Towards our beloved Earth, two good starting points are the Declaration on the Rights of Nature (or Mother Earth), taken by Indigenous Peoples meeting in Bolivia in April 2010, and recognition of Ecocide as a crime against humanity (see Higgins 2010).

DOI: 10.4324/9780429347481-9

Can we have real, effective dialogue about how we use our natural resources? I explored this question in the book *Ecology, Economy: Quest for a Socially Informed Connection* (Padel et al. 2013). How can we transform our economy from its present basis in ruthless competition and debt towards harmonization between economy and ecology, fair exchange with each other, and sharing our Earth's resources? The ideas of Bernard Cassen and others in ATTAC are important here, as are also the older ideas of Clifford Hugh Douglas concerning *Economic Democracy* (Douglas 1920), which—though effectively airbrushed out of economic history—have been shown to work in Canada, and depend on taxing *every financial transaction*, thus short circuiting the speculation and financial trading that drives scarcity and inflation. Douglas' main question is: can we create an economic system *not* based on debt?

Even more broadly, how can we enter dialogue in a way that bridges deep-seated conflicts and prevents war?

This book highlights the vital dialogue between Israelis and Palestinians promoted by "Roots" (in Chapter 2). We need dual narratives of how Israelis were killed and pushed out of their homes throughout Europe, North Africa, and the Middle East, alongside stories of the Nakba about how Palestinians have been displaced by Israeli settlers, starting in 1948, exactly when the UN Charter of Human Rights was released, and continuing in 2020.

We need similar dual narratives on Kashmir, by Kashmiri Pandits alongside mention of "disappearances" and stories of Muslim inhabitants who have lost loved ones to violence by security forces and others. We need similar dual narratives of the terrible ethnic cleansing that swept Greece and Turkey, resulting in the forced exchange of populations in the 1920s; in former Yugoslavia, and many other regions, until dialogue transforms "enemies" into friends everywhere.

Above all, we need to listen to the voices of Indigenous Peoples, starting to reverse the learning: they are our best guide on how to live with restraint in what we take from nature. "Taboo" is an English word taken from the Maori word *Tapu*, meaning "sacred": we need to recover a "sense of sacredness" in Nature (see Padel and Das 2010). This at a time when tribal boarding schools in India are being funded by the very mining corporations that are taking over and devastating tribal forests. This repeats the pattern of assimilation forced on Indigenous Peoples of North America, Australia, New Zealand, and other countries, for which the Prime Minister of Canada and others have apologized profusely. Just when these countries closed such boarding schools in the 1970s to 1980s, such boarding schools started to proliferate in India. Children's hair is cut short on enrolment and they are assigned an alien name in exactly the same way, whether from the Christian or Hindu tradition. Obviously, India's tribal people need education, but they need to exercise self-determination over the form this education takes. For example, tribal languages' exclusion from most schools is the reason for linguistic genocide throughout India (see Devy 2010).

If the next few years witness the epistemicide and cultural genocide of tribal or Indigenous communities in India, we lose the political and economic models of real, long-term sustainability that we need in order to survive as a human species. If this is true in India, it is also true in Indonesia, where the peoples of Irian Jaya or West Papua suffer terrible repression and displacement, in the Philippines, where Lumad community schools are being closed down, and in so many other countries, especially throughout the continent of America.

In the Middle East, Kurds have a special claim to indigeneity and offer models of similar value. *Kur-ti* is said to be Sumerian for "mountain people," and Kurds can be seen as resisting assimilation into nation states for 5,000 years, from Sumerian times to now. Rojava (north Syria) offers afresh new model of democracy, combining democratic confederalism with Jineology, questioning the nation state, and correcting the gender bias to ensure women's absolute equality in the power structure (Ocalan 2013).

But how to dismantle the world's wars, when the arms industry and trade are central to the world economy, and especially to the economy of every country termed "developed"? We call people "terrorists" who have been terrorized out of their homes by "security forces," when the greater terror is often this state terrorism perpetrated by governments. In this sense, double standards regarding Rule of Law are everywhere apparent: thousands of Adivasis and members of other marginalized communities languish in jail, though innocent, while policemen, soldiers, and corporate officials guilty of appalling crimes walk free with impunity (see Dandekar et al. 2014).

Our legal system is based on Roman law, and like our economic system and our standard model of democracy, is based on competition. The aim of Adivasi Law is to bring about reconciliation between contesting parties. The aim of Adivasi Economics is to exchange goods and labour equally and fairly. As for politics, as Jaipul Singh Munda said to Nehru during the Constituent Assembly Debates just after India's Independence: "How can you bring democracy to the tribal areas? Adivasis are the most democratic people on Earth. You have to learn democratic ways from them."

Can sharing as a core value—so evident in every tribal or Indigenous society—begin to replace the destructive value we have placed on competition?

Can we begin to meet in consensus, beyond the constant struggle between political parties, on Left and Right, and between egos in a system where power so often corrupts?

How can we transform our society and legal system to ensure that Environmental Justice is maintained? At present, we are far from this, and far from seeing how to make such a transformation. Thousands of well-intentioned people are working on different aspects—as this book affirms—but we need a new level of debate, rethinking core concepts about social life, decolonizing our methodologies (see Smith 2012), affirming (as the World Social Forum did) that history does not have to repeat itself: *a better world is possible.*

Recovering Justice From Irresponsibility[1] Towards Responsibility

Abdulkadir Osman Farah, Copenhagen University, Denmark

Much has changed since ancient times when justice mainly reflected how the powerful unilaterally thought uttered and acted. Despite certain improvements, the world still suffers from multiple transnational injustices resting mainly on power concentration and manipulation by the few. The nature, the environment, the economically marginalized, the ethnically distinct as well as other vulnerable groups in most societies incessantly search for balanced justices.

Drawing on numerous theoretical works and multiple empirical cases, combining the essence of justice with the dynamics of dialogical intersubjective human encounters and connections, Ananta Kumar Giri, in his latest work, *The Calling of Global Responsibility: New Initiatives in Justice, Dialogues and Planetary Realizations* calls for a new form of regenerative justice of global responsibility. Responsibility, not randomly combining justice and dialogue but insightfully weaving social, political, cultural, ethical, and aesthetical intra-group—as well as inter-group—connections. The approach modifies the social and political tendencies in which humans, historically as well as currently, often associate justices with power acquisition. Though justice remains a common social and political good, due to the persistence of human arrogance, and possible slumber, in prioritizing power centric hierarchies and systemic organization, eventually the wishes of the powerful might institutionalize and determine who will get what form of justice and who is potentially denied of justice. Consequently, under such controversial schemes and situations, the justice imposition from the powerful almost equates the justice ideals and practice of a given society.

For Giri, justice refers not merely to the structural criteria of producing universal social and political goods. Rather justice often occurs in a non-static dynamic process emerging, not from pre-given strategic framing, but situations in which people imagine, perceive, discuss, and intersubjectively project uni-dimensional as well as multi-dimensional reasoning. Considering it as transcendental "unfinished project," Giri refers to the analytical phrase "justice as responsibility." Such life-world process is characterized by continuing situated dynamic struggles. Struggles in which people strive for concrete living, respectful, and incorporating practices. The assumption

DOI: 10.4324/9780429347481-10

herewith includes that such situated social and political struggles should depart and build on progressive transnational dialogical, ethical, and aesthetical encounters and connections. From this perspective, "justice as responsibility" is neither given nor taken. Such socio-political practices ideally occur in "co-transforming" and coproducing social formations. This requires people moving beyond the currently restricted form of power centric justice. It might even demand expansion to "post-human constructions." Then, a kind of shared world could emerge. A world in which people, in their own terms and ends, deserve not just accession to the systemic designated institutional justices but also the active participation of ethically caring, listening, and dialoguing civic platforms. Ananta herewith recognizes that with such accommodating moral and spiritual approach, people could jointly counter and overcome the currently asymmetric and unfair globalized discourses often driven by intense dominating technology, commodification, and miscommunication. Such excluding discourses, often not grounded in the ethical civic sphere, remain detached from situated social conditions—creating widespread diverse forms of subordination and anxieties.

One ideal way of ensuring "justice as responsibility" requires the implementation of "restorative justice"—bringing together perpetrators and victims. In addressing injustices with public debates and other forms of inclusion might have worked in the past. In the current more complex, digitalized, and interconnected/interdependent world, the old nation-centric conceptions and approaches might represent a neglect rather than a nourishment. Ethics and aesthetics could instead help us in transformative ways in which the aesthetic dimension of social justice—dialogue between leading activists, communities as well as those they are struggling with or against, jointly fostering collective wisdom—meaning opening up their hearts and minds for others.

Most would welcome the conception and reflection on "justice as a responsibility" in an accommodating hermeneutic approach to social-political and cultural platforms. However, if we consider democracy as the political form of "justice as responsibility," the current political structures remain unbalanced in hierarchically organized democratic societies. Particularly those at the top (politically, socially, and economically) in their interactions with the majority of the society often frame dialogic, ethical, and aesthetic contexts. They do this mainly as means of improving their electoral as well as their socio-political power and legitimacy gains. Though this approach seems inclusive and accommodating within the so-called in-groups (majority)—it is far from dialogic, ethical, and aesthetic in relation to transnational ethnic communities—the so-called out-groups (others). Consequently, the politically powerful sees justice as the wishes and the priorities of the powerful/majority and often acts responsibility towards them. In contrast, in numerous occasions, the elites act irresponsibly in ignoring and occasionally oppressing alternative demands of justices.

On their part, in responding and overcoming such obstructive situations, ethnic minorities often seek alternatives in pursuing political, dialogic, ethical, and aesthetic engagements with diverse transnational allied and sympathetic civic constituents. In the short term, such encounters partially relieve community pain. In the long term, though, such steps fail leading to genuinely comprehensive social and political transformations. This is because such civic engagements have still to succeed in convincing the powerful in accepting more accommodation of others and thereby becoming more responsible.

Meanwhile, in practical terms, people, whether they are individuals, families, and communities, pursue a dignified life where justice remains integral in their daily routines. The struggles people involve might differ—but pursuing a fundamentally decent life in avoiding or resisting exclusions and oppressions represent human commonalities. For instance, people in many parts of the world as they pursue their daily routines also come into contact with authorities and officials responsible of exercising justice decisions in the society. In principle, as citizens, people bestowed with authority should also abide the common rules and adhere to the norms and jurisdictions of the society. Unfortunately, even in advanced democracies such as the Scandinavian context, such powerful individuals and groups often independently pursue their ego-centric political and social priorities. Occasionally, their political careers might demand in relating, for instance, to immigrants—not in a positive accommodating way—but in a negative derogatory manner. To avoid actual or potential troubles, many politicians prefer accommodating the unfair wishes and the priorities of the majority in the expense of the minorities. Politicians often engage continuing mutual dialogue and interaction with the powerful and the majority. In contrast, politicians pursue almost no dialogue and connection with the minorities. Instead, the political elite through the media negatively portray the transnational ethnic minorities—periodically profiling them as a threat to the society/majority.

Such powerful political groups often entertain new forms nationalism asserting superiority in claiming better humanity and sophistication compared to the underprivileged. According to Habermas, the current self-proclaimed nationalism should not be confused with patriotism or "constitutional patriotism" in which people inclusively and collectively promote responsible justices for all citizens—protecting for instance common democratic and dialogical values. What we instead witness today is the revival of vengeful nationhood through constant rivalry and power struggle and persecution. From such twisted discourses, the powerful sometimes attains approval from the majority both discursively and electorally. The process culminates in the presentation of legislations often curbing and violating the justices of already struggling minorities.

Meanwhile, minority constituencies struggle publicly and transnationally in mobilizing resources and resisting injustices in multiple ways. Occasionally media outlets and few experts refer to the suffering of certain minority groups—though the media often amplifies and further distorts the negative discourses and images generated by the politically powerful.

Eventually the disempowered active communities create their own transnational social and political platforms in which groups encounter and connect to other civic transnational social and political groups that may partially accommodate their concerns and potentially relieve their struggling conditions and situations. The following description of the work in Aarhus, Denmark, helps us understand this.

The Aarhus Case: About 10 years ago, in May 2010, in Aarhus, the second largest city in Denmark—and the Capital of Jutland—members of the Afro-Danish communities in the city established a grass root umbrella transnational civic association called Aarhusomali (www. aarhusomali.dk). The main aim of this association, bringing about 12 member-associations, was the consolidation of vibrant socio-political transnational community organization, the mobilization of community resources as well the contribution of development initiatives in the homelands some of the community members originate.

In the implementation of such efforts, the communities contacted authorities and established dialogical networks and roundtable meetings with politicians and civil servants. At the same time the community reached out diverse civic groups within and beyond Denmark. The transnational civic approach and engagement provided the community with additional capabilities the organization could not have generated through internal efforts alone. Among others this include the capacities of policy consideration and formation as well as the challenges of attracting relevant funding for diverse organizational activities. In addition, the communities also acquired inspiration and knowledge from established civic organizations, particularly the way in which these organizations strategize and deal with public authorities and powerful media organizations.

Consequently, while hierarchically reaching out politicians for accessing among others recognition, public goods and services, horizontally the communities mobilized members for better coordination and exchanges in empowering struggling community members at the different levels. Apart from efforts for better integration into the local and national environments, the communities also simultaneously connect to transnational civic developmental projects within Denmark and beyond. For instance, in the fall 2020, the communities schedule 4 major transnational debates in Aarhus—on the theme of "Black Lives Aarhus"—inviting politicians, civic advocacy groups, the media as well as the academia to discuss on the dynamics of transnational encounters and connections related to ethnic and race relations as well broader justice, democracy and development at local and transnational levels.

Justice is not just interpersonal or inter-group but also concerns intergenerational relations that remain the most vital/critical human relations. In the end therefore, what matters could be the transfer/transmission of inclusive knowledge and wisdom to future generations. If the current discourse maintained by the dominant populist-oriented elites prevails, then the world risks socializing future generations with "irresponsible justices." Otherwise if the powerful and the diverse transnational communities engage and prioritize microdialogues and moral virtues, societies could jointly transmit "justice as responsibility" resting on mutually beneficial knowledge and wisdom—to future generations. After all, we are all in one global village/vessel. The currently prevailing macro-structured world, where about eight powerful men unfairly accumulated the equivalence of more than half of humanity and earth's wealth, even the most powerful political, social, and economic elites might themselves already feel scared and disempowered. Regardless of status, we all have justices at stake. We must therefore imaginatively, thoughtfully, respectfully, and responsibly promote positive progressive transnational encountering and connections—that will among others—help us enhance "responsible justice" for and with mankind and beyond.

The United Nations Alliance of Civilizations and Searching for Improved Global Governance

Jeffrey Haynes, London Metropolitan University, UK

I have not known Ananta for long. I am however impressed by his knowledge and depth of understanding of the concept of global dialogue in pursuit of a better world. Ananta requested that I send him some of my writings on the Alliance of Civilisations (UNAOC), a United Nations's (UN's) body, created in 2005. I was happy to do so. He then invited me to add an afterword to his book. I am delighted to do so. Given that my writings on the UNAOC are my main contributions to debates on global dialogue, I offer here my thoughts about the UNAOC and, more generally, the UN and their contributions to global dialogue.

I became interested in the UNAOC, and by extension the UN, because of my long-standing interest in the role of religion in international relations (IR), a subject I have been researching into since the early 1990s (Haynes 2018a, 2018b). Many scholars pointed to a phenomenon: the "return" of religion to IR, which includes growing significance at the UN, characterized by a growing profile in both the General Assembly and the Security Council (Lehmann 2017; Carrette and Miall 2018). This was unexpected, following decades of virtual exclusion of religion at the UN. Over time, its importance has increased in relation to the wider issue of global governance.

In recent years, the United Nations has taken a palpable shift towards religious views, not least in relation to civil society, including faith-based organizations (FBOs). But how improved global governance can help solve international problems is a problematic issue, including how to achieve improved intercultural and inter-religious dialogue. Especially since September 11, 2001 (9/11), the focus has mainly been on intercultural conflicts between "Islam" and the West, although wider intercultural concerns—including relations between the West and Russia, China, India, and Iran—are also present. In response, the UN first created "Dialogue Among Civilizations" in 2001, followed by the Alliance of Civilizations in 2005. Beyond these specific initiatives, the UN is a focal point for numerous FBOs, which collectively seek to help bring about more international cooperation and less conflict in the context of developing "global governance" (Marchetti 2016).

Religion is a key component of any civilization. To understand the necessity of increased global dialogue, it is important to start with the role of

DOI: 10.4324/9780429347481-11

religion in today's international relations. Without improved inter-religious dialogue, better global dialogue is impossible. Beyond the UN, religion has greater significance in international relations, compared to a few decades ago (Sandal and Fox 2015; Thomas 2005). For some, the renewed significance of religion in this context is so great that it warrants the term "global religious resurgence." This has two main connotations. First, it implies a growing public voice for religion, in the sense that issues are increasingly viewed or framed through a religious lens. In addition, it is not only religious leaders and intellectuals, like the Catholic Church's Pope Francis, Archbishop Justin Welby of the Anglican Church, and the Muslim intellectual, Professor Tariq Ramadan, who publicly express concern with the relationship between religion and economic and social justice. In addition, around the world, numerous FBO leaders proclaim their desire to help make societies more just, more equal, and more focused on spiritual issues. They use a variety of tactics and methods in pursuit of their objectives. Some, like the Anglican Church in Britain, lobby, protest, and publish reports at the level of civil society. Others use the opportunities afforded by high-level meetings to pursue their objectives. Overall, what encourages religious leaders to voice their social and economic concerns? Berger maintains that what they have in common is a critique of secularity, because human "existence bereft of transcendence is an impoverished and finally untenable condition." He argues that a human desire for transcendence—that is, a state of being or existence above and beyond the limits of material experience—is an integral part of the human psyche, and secularity—that is, the condition or quality of being secular—does not allow for this necessary sense of transcendence. Without a sense of transcendence, Berger asserts, life for many people is unsatisfactorily empty (Berger 1999: 4).

The "return" of religion to international relations underlines that the Enlightenment belief was wrong: that all societies would inevitably secularize along identical or closely related linear pathways, leading to uniformly similar ways of "modernizing." The combined impact of modernization (involving urbanization, industrialization, and swift technological developments) and globalization, with capacity to transmit capitalism around the world with, inevitably, winners and losers, coupled with a growing lack of faith in secular ideologies (communism, socialism, and liberal democracy), has left many people with strong feelings of alienation and loss. The result is a (re)turn to seeking transcendence, typically in various religious vehicles, although not necessarily associated with traditionally dominant faiths, which in my home region Western Europe, would be the Catholic and various long-established Protestant (Anglican, Lutheran, etc.) churches. Undermining "traditional" value systems and allocating opportunities in highly unequal ways within and among nations, secularization has helped produce in many people a deep sense of disaffection, stimulating a search for a clearer identity to give life meaning and purpose; many people find what they want in various faith expressions, including grassroots religious movements. In addition, the rise of global consumerist culture has led to expressions of aversion, sometimes focused in the concerns of faith groups. Overall, the result is a wave of resurgent faith—with far-reaching implications for social integration, political stability, and, in some cases, regional and international peace and security. Religious resurgence occurs in a variety of countries with differing political and ideological systems, at various levels of economic development, and with diverse religious traditions. All have been subject to the destabilizing pressures of state-directed pursuit of modernization and secularization; in other words, all are experiencing to some degree the postmodern condition—characterized by a lack of clarity and certainty about society's current and future direction.

Resurgent religion does not only relate to personal beliefs. It can also lead to a desire in both individuals and groups to seek to grapple with interlinked social, economic, and political issues. "Because it is so reliable a source of emotion," Tarrow (1998: 112) remarks, "religion is a recurring source of social movement framing. Religion provides ready-made symbols, rituals, and solidarities that can be accessed and appropriated by movement leaders." They are found in many different faiths and sects, sharing a desire to change domestic, and in many cases international, arrangements: to "do good" or "do better" by projecting the influence of their religious faith into this-world action. Religious actors adopt various tactics to try to achieve their goals, at various levels of organization. Some protest, lobby, or otherwise engage with decision-makers at home or abroad. Others focus reform intentions through the ballot box or via civil society. Still others—a tiny minority—may even resort to political violence or terrorism to try to pursue their objectives. Finally, numerous religious actors of various kinds seek to engage in current political, economic, and social debates, including at the United Nations and other global and regional fora (Haynes 2016).

Many scholars and commentators have noted religion's destabilizing effects, with numerous references to the notorious events of 9/11, which took place two decades ago (Haynes 2018a). This was of course the day when 19 al Qaeda terrorists attacked the USA, "armed" "only" with aeroplanes, resulting in the deaths of 2,996 people. Second, more generally, there is the impact of globalization, highlighting a need for improved "global governance" to deal with myriad problems, including the rise of religion-linked terrorism and extremism. Yet, despite much emphasis on the destructive, conflictual elements of religion destabilizing global politics, its overall international impact is not easily interpreted (Sandal and Fox 2015). On the one hand, inter- and intra-religion conflicts in many regions and countries are a significant source of domestic and international strife, attracting much media attention. On the other hand, it is often suggested that if "benign" and "cooperative" religious principles and practices could be consistently applied to try to reduce conflicts then what Falk (2004: 137) calls "emancipatory religious and spiritual perspectives in world order thinking and practice" might improve at least some outcomes in global politics. For Falk, what is needed is a shift to what he calls "humane global governance," that is, an improved moral and ethical regime, informed by both religious and secular insights.

Difficulties in coming up with adequate responses to issues of global concern frequently result in criticism of the world's only universal inter-governmental organization, the UN (Puchala et al. 2007). Founded in 1945, the UN is a state-centric entity primarily concerned with international secu-rity and cooperation. For the UN, 9/11 was a profound challenge, reflecting a new focus of conflict in international relations—inter-cultural/inter-religious conflict, one with which the UN was not obviously equipped to deal, but nevertheless was expected to do so. At the UN after 9/11, there were fre-quent references, including in the General Assembly and Security Council, not only to religion's perceived links to extremism and conflict but also to the necessity of improving inter-faith dialogue as a means to improve global dialogue more generally (Lehmann 2017).

What is the likelihood that the United Nations from its creation in 1945 an emphatically secular, state-centric organization is equipped to offer reli-gion a sustained voice, to the same extent as secular approaches, in the search for improved global governance? At first glance, this seems unlikely, given that historically the UN has shown indifference at best and antipathy at worst to religion (Haynes 2014). While there is a mention of religion in the founding UN Charter, this is a passing reference only—in the context of freedom of belief as a fundamental human right (Haynes 2018a). On the other hand, in recent years, the UN's focus on human rights has become an area of dispute, with different sets of values in competition in what Puchala et al. (2007: 128–135) call "the politics of culture." In relation to the latter, the UN now acknowledges a direct link between development shortfalls and the rise of international terrorism and violent extremism. This extends the UN's concern with human rights to an understanding that for human rights to be improved for all would require immediate and comprehensive action to deal with endemic development shortfalls in the world's poorest countries. This was reflected in the Millennium Development Goals (MDGs; 2000–2015) and their successor, the Sustainable Development Goals (SDGs; 2015–2030). The SDGs have already achieved iconic status at the UN, and much of its work, across its dozens of specialized agencies, is focused upon achieving the ambitious goals of the SDGs.

Adhesion to the SDGs and Agenda 2030 reflects the UN's recognition that to achieve improved, more equitable, international outcomes, for example, in relation to poverty-alleviation and development, it needs to do more than supply rhetoric moral and ethical support. Why is this? As Lynch (2012) states, the UN focuses on international development shortfalls, moving over time from initially moral dimensions to consider material factors: "neolib-eral competition of the 'market' [in] international development." From there, as she notes, it is but a short intellectual jump to consider how globalization appears to encourage or exacerbate an already unjust and polarized world, where the rich benefit disproportionately. This reflects the fact that the past three decades—that is, the post-Cold War era—was a time of deepening globalization and societal and political polarization. This coincided with religion's increased international impact, overlapping with ethical and moral concerns regarding global governance. However, as Puchala et al. (2007: 117) note, "questions of values [at the UN] get politically transformed into issues of hegemony, autonomy, privilege, and power." This is a way of saying that such concerns at the UN cannot plausibly be assessed in isolation from how the organization is structured and how member states wield their power. There are no formal ways in which non-state actors can affect UN decisions and policies. However, as already noted, this does not stop religious views and opinions being regularly expressed—not least by the hundreds of FBOs registered with its Economic and Social Council. They are concerned with the nature and impact of post-Cold War globalization, in relation to: international development, "climate change, global finance, disarmament, inequality, pan-epidemics and human rights" (Carrette and Miall 2013: 3). As a result, many UN debates are increasingly "by the moral resources that 'religions' offer and agencies of global governance need an awareness of what religious actors are doing and sensitivity to religious difference" (*ibid.*). In short, post-Cold War deepening and widening of globalization has paved the way for a highly unexpected development: greatly increased religious and spiritual energies in international relations, a development with variable impacts.

The UN is now aware of the power of religion to move individuals and groups to action, in both normatively "good" and "bad" ways. Chiming with a growing understanding that the UN was too "top down" in orientation in the 1990s, too remote from the concerns of "ordinary people," the UN became concerned with interacting with "civil society," both secular and religious manifestations, seen as necessary in trying to improve global governance and outcomes. The World Conference on Religion and Peace (sometimes referred to as the "UN of Religions") asserted:

> [R]eligious communities are, without question, the largest and best-organized civil institutions in the world today, claiming the allegiance of billions of believers and bridging the divides of race, class and nationality. They are uniquely equipped to meet the challenges of our time: resolving conflicts, caring for the sick and needy, promoting peaceful co-existence among all peoples.
>
> (World Conference on Religion and Peace 2001;
> no page number supplied, quoted in Berger 2003: 2)

Religion's increased impact stimulated an understanding that today's international relations are not exclusively secular. Instead international relations are now significantly affected by religious norms, beliefs and values, leading to what some identify as "postsecular" international relations. Reflecting these concerns, the UN has moved over time, via the UNAOC, to express consistent concern about international religious freedom.

Reflecting this change, in recent years, global public policy debates and discussions at the UN have undergone a shift in emphasis from exclusively secular and material to include moral and ethical issues, which frequently overlap with faith-based concerns, for example, in relation to the MDGs and SDGs.

Compared to the UN's early years, the organization is now replete with religion-influenced activities, with many opportunities for the various voices of religion to make themselves heard. These include the activities of a "religious state," the Holy See, as well as those of the world's second biggest inter-governmental organization, the Organisation of Islamic Cooperation, and the aforementioned hundreds of registered FBOs. Religious entities have not adapted to the UN's secular culture merely to survive—but to develop their own voice.

Finally, what of the UNAOC after 15 years? It would be impossible to claim that it has clearly made a sustained and palpable improvement to global dialogue (Haynes 2018b). The world in 2020 seems as hell bent as ever in not dealing with the profound problems of our age—the climate emergency, attacks on human rights, the widening gap between rich and poor, the decline of democracy, and so on. Does this point to a fundamental problem with our governments: they are simply not equipped to work collectively, whether at the UN or elsewhere, to devise the policies and programmes necessary to transform our world for the better? Would bringing FBOs and other religious entities more centrally into the discussions help or hinder the process of finding a way forward? Ananta's book investigates these and other issues with great aplomb. He points to the potential of people to make things better. The key question now is: how can this vital task be accomplished?

What is my perspective on dialogue and global responsibility and what are the challenges facing us all in these respects? The planet is beset with very significant challenges. It is invidious to try to place them in a hierarchy of concern. From my perspective, however, the immense challenge of the climate emergency stands out as both the most imminent challenge and the one in which we all have key responsibility. We have personal responsibility to treat the planet better, and we also have a collective responsibility. We must pressurize our governments to take the climate emergency seriously and to devise and operationalize policies—both individually and collectively—to tackle the problem. We don't have long to do this! There is also the serious problem of global inequalities: low wages, inadequate health, and safety provision at work, and difficulties of achieving an acceptable work–family balance. Once again, to act individually in trying to diminish global inequalities—via the products we buy and the choices we make about where we holiday and so on—is necessary. It is also to pressurize on our governments to address these concerns both at home and in the global context. In short, the answer to global problems lies both in greater individual responsibilities and in the willingness of governments to do the right thing and work towards sustainable and more equal global outcomes.'

International Charter of Responsibility

The Work of Charles Leopold Mayer Foundation and Beyond

Julie M. Geredien, Independent Scholar, Annapolis, USA

In his book, *The Calling of Global Responsibility: New Initiatives in Justice, Dialogues and Planetary Realizations*, Ananta Kumar Giri brings attention to different movements and initiatives that respond to the call for global responsibility. An important theme in his discussions of these has been the need to open a global dialogue that will cultivate understanding of the "multiple meanings and languages" associated with the term responsibility. In Chapter 3, which investigates new initiatives in dialogues across borders, he introduces the International Charter of Responsibility and highlights the work of Edith Sizoo and the Charles Leopold Mayer Foundation for Human Progress. In the following pages, I offer the reader a more detailed account of the efforts involved in this particular initiative, and describe some related wider issues of theory and practice.

The Charles Léopold Mayer Foundation for Human Progress (formerly "Fondation pour le Progrès de l'Homme") was founded in 1982 under Swiss law. Its goals are to support the emergence of global community by focusing on governance, ethics, and sustainable models for living and development. Charles Leopold Mayer was born in the latter part of the 19th century and was a chemist and engineer, though he made his fortune in finance. He wrote works on chemistry, spirituality, and humanism and was inspired by Alfred Nobel a Swedish chemist, engineer, and armaments manufacturer, about 50 years Mayer's senior, who was famous for his invention of dynamite. Having read a "premature obituary" that criticized him for selfishly benefitting from the sale of cannon and armaments, Nobel dedicated his fortune to establish the Nobel Prizes in 1895. Charles Leopold Mayer supported the progressive values that were honoured through these annual awards and similarly invested his own wealth to upraise scientific and humanist ideals. When he died in 1971, Madeleine Calame, his secretary for 30 years became the executor and administrator of the proposed foundation which now bears his name.

The Charles Léopold Mayer Foundation for Human Progress does not limit its endeavours on behalf of global community to the academic world but instead has engaged in long-term partnership with public and private partners and a very diverse range of social and occupational groups, defining

DOI: 10.4324/9780429347481-12

itself as a "human adventure." It views itself as a learning organization, dedicated to the value both of particular initiative and synergistic linking of initiatives to develop new insights for humanity. Through donations and loans, it seeks to finance research projects and to coordinate initiatives that support the "significant and innovative progress of mankind."

Since 2003, Edith Sizoo, a socio-linguist working within the framework of Development Cooperation, has coordinated an international process that was initiated through The Charles Léopold Mayer Foundation for Human Progress, to promote a Universal Charter of Human Responsibilities. The Charter can be understood as an initial embodiment of an "ethical" third pillar required to progress international relations in the 21st century. The two established pillars are the Charter of the United Nations (*UN Charter*) created in 1945 as the foundational treaty of the United Nations, focusing on peace and development, and the Universal Declaration of Human Rights, created in 1948 to defend the dignity and rights of individuals. These two pillars have provided a framework through which organized international relations can, in some regards, be said to have progressed in the 20th century. However, Sizoo, and countless more like-minded souls, believe that in the past 50 plus years, radical global changes and multiple crises have made the rise of a third pillar a necessity for optimal peaceful and sustainable global development. This pillar focuses humanity on an ethic of responsibility that moves beyond an anthropocentric and merely socio-centric framework for law and human relations.

The need for a third pillar was first seriously publicly discussed at the 1972 Stockholm World Conference, when the idea was presented to establish an "Earth Charter," which would focus mainly on responsible relations between humankind and the biosphere. This idea was revitalized during preparation for the 1992 Earth Summit in Rio de Janeiro, although it was not until 1994, that a six-year worldwide consultation process was called together by Maurice Strong and Mikhail Gorbachev with the purpose of developing "a global consensus on values and principles for a sustainable future." The preamble of the UN Earth Charter, which locates humanity at a "critical moment in Earth's history," clearly establishes the "third pillar" concept of universal responsibility, stating that "it is imperative that we, the peoples of Earth, declare our responsibility to one another, to the greater community of life, and to future generations," and also that "we decide to live with a sense of universal responsibility, identifying ourselves with the whole Earth community as well as our local communities." The Earth Charter further recognizes the emergence of a global civil society as a new frontier requiring that we recognize environmental, economic, political, social, and spiritual challenges as being interconnected and as requiring "inclusive solutions."

However, the initiative to establish a third pillar for international relations, grounded in the ethic of responsibility, took several different additional forms during this general time period, all of which have influenced the international process coordinated now by Sizoo through The Charles

Léopold Mayer Foundation for Human Progress to develop and promote the Universal Charter of Human Responsibilities. For example, the Alliance for a Responsible, Plural and United World, was initiated in 1993, a year before consultation officially began on the Earth Charter, when the Alliance published its founding document, titled "Platform for a World of Responsibility and Solidarity." This text calls for a unification across the divides that presently still separate South and North, poor and rich, men and women, nature and humankind. It was able to mobilize people across the world to begin framing proposals that protect human dignity and all planetary life. Similar like-minded initiatives that have interacted with the Alliance are the *Declaration towards a Global Ethics*, created by the Parliament of the World's Religions in Chicago in 1994; the *Universal Ethics Project*, prepared by the UNESCEO Division of Philosophy and Ethics; and the *Universal Declaration for Human Responsibility*, written in Vienna in 1997 by the Interaction Council Congress.

The Universal Charter of Human Responsibilities has from the outset been an evolving text, as it is continually revised and adapted in response to consensus-building processes. Workshops in Africa, Asia, Latin America, and Europe that looked to the daily realities of life in different societies so as to find guiding common values and principles from among them were organized from 1995 to 1998 by Andre Levesque and a cohort of friends. By 1999, a first draft of the Charter had emerged from these wide-ranging efforts. Then, from 1999 to the end of 2000, the draft Charter was assessed to determine its particular applicability within different cultural contexts and fields of human engagement. During this time, additional Alliance groups worked within their fields to develop proposals to address the diverse challenges of the 21st century. Based on these assessments and proposals, a decision was made in 2001 to prepare a new draft of the text that was inclusive of all these inputs and insights. A Committee of Wise Persons was given the text in the fall of 2001 and was able to amend and revise the initial draft. The revised draft was then given in the winter of 2001 to the World Assembly of Citizens, a group in Lille, France that was organized by the Alliance. Participants related the draft directly to their own diverse background experiences and submitted comments, which led to a further revised text, which, after the Lille Assembly, was submitted to the Alliance for further comment, leading to a text that began to be widely disseminated starting in October 2002.

In 2003, Sizoo, who is currently the International Coordinator of the Forum on Ethics and Responsibility, began her efforts through The Charles Léopold Mayer Foundation for Human Progress, to coordinate an international process that will promote cultures of responsibility and the idea of a universal charter that outlines human responsibilities for all peoples on the planet and serves as the necessary ethical third pillar to unify and mobilize planetary realizations and international relations in the 21st century. Sizoo's strong feeling about the need for a universal charter addressing human responsibilities continues to resonate today with the mid-20th-century political, intellectual, and spiritual realization that the crisis of Nazism and the larger global predicament of systemic dehumanization requires a pro-active and unified moral response from surrounding nation-states.

Sizoo grew up in the Netherlands during World War II and experienced in her youth war time scarcities in food, shelter, and clothing as well as reports of atrocities towards Jews. In an interview published in 2013, she said,

> Witnessing and experiencing the events of the war, I was driven by questions such as, what is racism and its causes? . . . This led me to go as deeply as possible into learning to understand what this otherness consists of so as to diminish the fear and to discover the richness of human and cultural diversity.
>
> (originally published October 10, 2013 on Newzfirst by Shoaib Mohammed, www.newsfirst.com/web/guest/)

Sizoo's background therefore informs her present-day insistence that we cannot stand aside and do nothing in the face of evil. All peoples are called both uniquely and commonly to take responsible action to address the global environmental, economic, social, political, and spiritual emergencies of our day and to engage our gifts of criticality and caring to overcome all forms of human hatred, reactive fear, and exclusion.

An International Charter Facilitation Team associated with The Charles Léopold Mayer Foundation for Human Progress involves itself with groups and organizations in approximately 25 different countries around the world and engages in this international process coordinated by Sizoo. The effort has been funded by the Charles Leopold Mayer Foundation, while on the local level, activities are financially supported by a range of local organizations as well as voluntary contributions. The process of developing and promoting the charter is understood to unfold through continuous dialogue. Although the national Charter committees and their partners are mostly involved in civil society and its organizations, their aim is to "build bridges" between civil society and influential decision-makers within political and economic domains of life. They seek to differentiate and integrate distinct dimensions within the public sphere, including social institutions, markets, scientific institutions, international economic institutions, businesses, and religious institutions. Those participating in the process include women, young people, people living in precarious conditions, artists and publishers, trade unionists, theologians and philosophers, teachers, non-governmental organizations, politicians, business people and engineers, journalists, civil servants, scientific professionals, people of military and juridical backgrounds, and people related to the health sector.

The initiative views itself as a worldwide movement and seeks not only to create cultures of responsibility and to promote the charter but also to connect the idea of responsibility to actual concrete issues and practices experienced in daily life. It perceives responsibility as a mentality, which also reflects the moral and spiritual cultivation of a particular emotional and cognitive disposition. In this way, it promotes an appreciation of ontology and an understanding of praxis and reflective action. Furthermore, it recognizes a groundswell of initiatives that are searching for models to address the current critical threats and issues affecting all life on the planet and aspires to further mobilize these efforts so that they reach a collective magnitude and visibility which obliges the world of international politics to address their claims. Two advisory groups of "experts" were formed to examine feasible and "efficient" ways to reach their international political objectives and there is continued search for other like-minded international initiatives, in addition to ongoing efforts to communicate with national governments, high-level political personalities, and intermediate UN institutions.

The activities of the International Charter Facilitation Team aim then for both an inner and outer, bottom-up and top-down, approach: they aim to reach high international political levels and also to affect the everyday reflective life practices of diversely culturally identified citizens around the world. It is also important to note that the International Charter Facilitation Team understands the charter to be a starting ground and a "pre-text," or tool, for dialogue. It does not view the charter as a finalized document waiting for a stamp of approval from those on either the grassroots or high governmental level, but rather, as a catalyst that will help to bring forward the necessary insights for new models of development that support and sustain our collective evolutionary progress.

In a speech offered by Qin Hui at the closing 2001 session of the World Citizens Lille Assembly, organized by the Charles Léopold Mayer Foundation for Human Progress, Hui speaks on the topic of cultural multi-identity, touching on themes of diversity in unity and harmony-making which remain central today to this worldwide movement for a universal charter. Hui understands the movement for a charter to be the expression of a plural and humane globalization that prioritizes the common good, responsibility, solidarity, and principles of balance, and that is rising as a counter to a purely economic globalization of homogenization and dehumanization, based on market competition.[2]

To Hui, the charter itself is an ongoing work in progress that does not replace the 1948 Universal Declaration of Human Rights but rather seeks to complete it. In the plural and humane globalizing movement that continues to advance the people of the earth towards this complementation and completion, Hui stresses how important recognition of multiple identity is, as it makes possible a much more full realization of the right to choose one's cultural identity, and on a deeper level, provides humanity with epistemic access to the plurivocity of the human self (Giri 2013). This realization is critical for maturation processes related to responsibilization. Hui says:

> If cultural pluralism meant that "Everyone can only identify with their native culture," it would make the "trans-cultural", who lives with several cultures, "marginal" and discriminated against everywhere.

Therefore, I suggest that the Charter emphasizes the importance of respecting cultural diversity and that it promotes tolerance between cultures, but also that it recognizes the importance of multi-culturalism, born of the meeting of cultures since ancient times. As we do not accept the idea of cultural hierarchy, we cannot accept that a "pure" culture is superior to "mestizo" cultures. This is also why any "movement of purification" imposed within a cultural area cannot be legitimate.

Ultimately, the recognition of the multi-identity is a new dimension of human rights. People have the right to choose cultural identity and the responsibility to respect the cultural choice of each individual. A true multi-culturalism opposes any kind of cultural constraint: both forced cultural assimilation and "anti-assimilation". The Charter has made a step forward by focusing on respect for diversity. However, we have to further develop the dimension of the right to multi-identity.

In a report on that World Assembly of Citizens gathering in Lille France, Sizoo addresses a problematic issue at the heart of Hui's concern. She reports:

[T]here were many problems with the meaning of key-notions given the wide variety of cultural contexts people come from. For instance, our Asian friends had problems with the notion of "unity" in diversity, saying that "unity" is a hegemonistic concept and that we should rather speak of "harmony". The notion of "responsibility" itself, of "justice", or "peace" need to be contextualized as well.

Interestingly, in ancient Chinese understandings of harmony (ho) and the virtue of yi, there are two senses in which one may relate to the concept of unity; by articulating the difference between these two kinds of unity, one approaches a conceptual framework for addressing the problem reported by Sizoo and implied by Hui in the speech given at the close of the assembly. In one sense, unity may manifest as conformity, "a static state of oneness without adaptability" (Cheng). In this one-dimensional expression of unity, differences are transcended but they are never integrated. However, in another sense, unity can be understood as creative and productive of both differences and new levels of relationship and synergy. When a discerning and responsive form of human reason, or yi, is engaged, people are able to distinguish differences in parts "without losing the unifying ground and the unifying principle" through which their commonality is felt and cognitively understood. They are able to maintain the perception of unity simultaneously with the perception of difference, even under changing circumstances.

This second kind of creative and awakened unity "is the basis for producing and strengthening harmony as a system of integration of differences" (Cheng 1989). Freire speaks of this quality of discernment that is able both to distinguish and to integrate differences as *conscientização*. *Conscientização*, or critical consciousness, arises from an over-flowing reverence for life, called *Ehrfurcht vor dem Leben* by Albert Schweitzer, and referred to by Freire as bio-philia. Sizoo herself speaks of these foundational qualities of being in her own way, as: "a passion of love for all that lives, other human beings, flowers, animals, the beauty of nature." Both a sense of discernment and a sense of responsibility arise from deep within us, she believes, "because of our love for this immense creative force of Life itself." In addition, she describes a kind of perseverance and determination that is also essential in the process of creating integrated relationships through the perception of difference. In her report on the 2001 World Assembly of Citizens meeting, she further writes: "'Love' originally means 'staying with'. Taking responsibility out of love then implies 'staying with . . . ', that is: not giving up, being aware, being awake and acting from love for all that lives."

Another helpful insight about unity to consider here emerges in the Neo-Mohist Canons. In this text, there is a comprehensive view offered of the Chinese concept of identity (*t'ung*). While the act of established an identity activates the "categorizing mind" that characterizes formal operational rationality, the Neo-Mohist Canons offer understandings that allow one to establish identity without necessarily partaking in modernity's reductionist materialism or in limited qualities of mind often associated with acts of identification, that are influenced by fear and ignorance, such as racism, parochialism or anthropocentrism. Cheng summarizes content on this from the Canons:

> [F]our kinds of identity are distinguished, and the concept of identity (*t'ung*) is thus defined for four different kinds of identity: the identity of one thing referred to by two names; the identity of belonging to the same body; the identity of grouping together in the same space; and the identity of belonging to the same class (lei). Given these distinctions, it is evident that not all identities are exclusive of differences on whichever level. In fact, *identity could become a harmony of differences without negating or eliminating the differences.*
>
> (Cheng 1989: 234, emphases added)

These insights about the nature of identity are practical for those involved in this worldwide 21st-century movement, because for them, responsibility is more than a personal ethical principle. Instead, it is a commitment made by citizens who share a common social identity. The initiative to compose a Charter of Human Responsibilities therefore obliges those involved to participate in an exploration of the common values and principles that underpin this identity. It additionally obliges them to find a way to express this shared identity in a way that does not negate Hui's plea for multi-identity and trans-cultural identity.

Sizoo, in referring to the realization of diversity within the universal ideas and principles that relate us commonly to the understanding of human responsibilities, adopts a "geo-philosophical" mode of thinking in order to fulfil these obligations, in which the earth is not just one element among other elements but the encompassing source of all elements. In the afore-mentioned 2001 Lille report, she insists that: "we cannot let it happen that our children and grandchildren grow up in a world where . . . the respect for Mother Earth as a living organ" is not practiced as an essential guiding principle for behaviour and in which "patriarchal domination and values continue to prevail in our societies." These moral imperatives imply that human beings are part of a "woven" universe, of inter-connected realities and inter-relationships. Integral human development and holistic well-being are inseparable from the sustainable and flourishing well-being of all earth systems. On this path of "responsibilization," the egoic efforts that perpetu-ate individualism are replaced with the relational and altruistic labours that birth individuation, fostering genuine evolutionary human growth through integrity to principles that uplift our shared humanity.

To appreciate the primordial significance of Sizoo's geo-philosophical assertions, and how they relate to a matriarchal transformation towards a responsibility framework for law and social body order, it is helpful to recall one of the four kinds of identity *(t'ung)* established in the Neo-Mohist Canons summarized earlier: that is, the "identity of belonging to the same body." Cheng affirms, "In particular, the identity of belonging to the same body is quite compatible with the notion of *ho*, [harmony] and indeed basically exemplifies the same structure of *ho*: integration of differences as parts in a unity of totality" (p. 234). At the 2001 Assembly, Sizoo presented just this kind of a way to forge a shared identity. She offered to people an image of the Charter in the form of a marguerite with a heart and a large number of multi-coloured petals. In the common heart of this flower, abide all of the commonly agreed principles and values pertaining to human responsibility, including right to a life of dignity and respect for non-human forms of life, a preference for dialogue rather than violence, compassion and consideration for others, solidarity and hospitality, truthfulness and sincerity, peace and harmony, justice and equity, and a preference for the common good rather than self-interest. The more outward petals extending from this heart would represent the diverse applications of these principles and values, in socio-professional fields, as well as culturally adapted translations for the various linguistic contexts.

Another metaphor, presented by participants striving to establish the char-ter, communicates how guiding principles may serve as a common nucleus, and how they may be transferred and adapted into different fields of human endeavour through translation into culturally appropriate forms. This meta-phor asks people to see the "common ground" of shared principles as "the roots of a tree like the banyan." The roots, or principles, produce a large number of branches and new trunks, which denote the application of the common guiding principles in a variety of cultural contexts and fields of human activity. In both examples, one can see how recognition of the need to care for the larger whole of cultural context and planet connects us to common and central harmonizing principles of caring, from which new responsible forms of social life may develop and grow. Both the metaphors of flower and tree, and the general recognition of Earth as inter-related living organism, relate people to a sense of identity through the recognition of "belonging to the same body." In this form of identity making, integration of differences becomes participation in the inclusive unity of a vast totality. The awakening of consciousness to shared human responsibility therefore becomes an expression of the natural law that is infused in the whole, as it allows us to realize our natural rights and to develop optimally and providentially within every context of life on the planet.

Ten principles offered in the Charter that presently guide the exercise of human responsibility are: responsibility as an aspect of constituent power, related to inherent moral and spiritual power of all human beings; participatory nature of dignity and freedom making; respect for optimal development of humanity, including material and non-material aspirations; unity in means and ends, process and outcome; principled and precautionary thinking; ethical criteria for scientific investigations and research; accountable use of power to serve the common good; practice of high reason; ongoing discernment of and respect for cultural specificities. Additional principles that may hopefully soon be further considered for inclusion in the "heart" of the responsibility flower and at the "root" of the responsibility tree are: concern for continuous re-centring of those peoples and public issues abiding in margins; cultivation of ethical disposition and qualities of "good mind" such as impartiality or non-bias; respect for health and integral well-being on all levels, especially as related to the relationship between land, agricultural practices and food; embrace of practices of remembrance, including collective mourning, holy and historical days, respect for ancestors and compassion and gratitude for past human suffering; development of new epistemologies and modes of discourse that evolve human communication across all borders; appreciation of the need for role models and mentor practices in human development and for new human roles of stewardship and trusteeship; embrace of genuine life-long learning, engaging all forms of media and universal public access to education.

Sizoo and fellow reflective activists understand that responsible action requires that different categories of human activity be integrated. Each of the principles and values reviewed earlier is inter-connected and cannot be addressed in isolation. The harmonizing principles and values at the "heart" of the Charter are considered absolute orientation points, from which all social and professional spheres may draw up their own guidelines for responsibilities. These guidelines, once composed, serve as the foundation of a social agreement that connects each sector of society to each other and allows for the emergence of a global civil society. This foundation of agreement makes possible new realizations of human dignity. As Jacques Maritain, the key philosopher behind the Universal Declaration of Human Rights has written, "A person possesses absolute dignity because he is in direct relationship with the absolute. His spiritual fatherland consists of the entire order of things which have absolute value" (Maritain 1944). In the global civil society that the Charles Léopold Mayer Foundation for Human Progress aspires to help bring forth, all citizens may realize a form of "absolute" human dignity, by recognizing the internal togetherness, or unity, of the larger whole of which they are an invaluable and inseparable part. This liberated form of identity can only be achieved through acceptance of "absolute" orienting values and principles that conduce to the well-being of all humanity and that safeguard its interests. The process involved embodies the emergence of a new worldwide consciousness which will be a veritable evolutionary benchmark for humankind. The ethical third pillar for global law and relations will reflect this remarkable dawning of consciousness.

The Calling of Global Responsibility

Corona Crisis, the Israeli–Palestinian Conflict, and Rays of Hope From the Congress of the People

Sapir Handelman, Minds of Peace, Tel Aviv, Israel

In his book, *The Calling of Global Responsibility: New Initiatives in Justice, Dialogues and Planetary Realizations*, Ananta Kumar Giri discusses many initiatives in justice and dialogues. One of these is the work of Roots, a voluntary organization working for peace and transformation of conflict in Israel and Palestine. I have been working on transformation of Israeli Palestinian conflicts now over a decade. But the present Corona crisis distracts political and public attention from major problems, such as the Israeli–Palestinian conflict. These problems will not suddenly vanish or spontaneously be resolved. It seems inevitable to tailor a "constructive" link between the two crises: Can a critical discussion on the optimal way to cope with the Corona crisis create peacemaking opportunities? Can peacemaking initiatives have the potential to improve the ability of Israelis and Palestinians to cope with the Corona crisis?

The last meeting of the founding members of the Congress of the Israeli–Palestinian People—a grassroots initiative—linked between the two crises. The groups looked at the Corona crisis as a case study to examine the advantages and disadvantages of three alternative solutions to the conflict: Two State solution, Federation, and Confederation. The critical discussion in the assembly illuminated significant aspects in the challenge of change and demonstrated the importance of public involvement in the struggle to cope with major social problems.

The link between global pandemic and communal conflict with global implications concretizes the interplay between public, leadership, and global responsibility, which is one of the main focuses of this book.

A Revolutionary Peacemaking Process: Leaders, People, and Institutions

The Israeli–Palestinian conflict is a classic case of intractable conflict. It is a long-time struggle where generations in turn are born into reality of violence and aggression. The conflict seems to have a life of its own. It seems to operate by a destructive evolutionary mechanism: almost every element that benefits the conflict survives and whatever works against it becomes extinct

DOI: 10.4324/9780429347481-13

or insignificant.[3] A revolutionary peacemaking process is required in order to break the chain of destruction.

Three critical elements are necessary to initiate a revolutionary peace process: visionary leaders, public involvement in the struggle for change and peacemaking institutions (see Handelman 2012a). These elements are intertwined. Visionary leaders need public support to initiate an effective peacemaking policy.[4] People that are involved in the peacemaking struggle invite visionary leaders to the political stage and demand that they initiate a peacemaking policy. Peacemaking institutions connect visionary leaders to the public and operate a revolutionary peacemaking process.

Peacemaking institutions operate in two opposite directions. On the one hand, peacemaking institutions are political tools for visionary leaders to generate public support in the peacemaking. On the other hand, peacemaking institutions are political instruments for the people to motivate leaders to negotiate peace and conclude agreements (compare to Handelman 2012b). The Public Negotiating Congress is an example of a peacemaking institution.

Public Negotiating Congress invites representatives of the opposing people to discuss, debate, and negotiate solutions to the conflict by peaceful means. The idea is inspired by the multi-party talks of the 1990s that helped to create a revolutionary change in two difficult cases of intractable conflict: the "trouble" in Northern Ireland and the struggle against Apartheid in South Africa.[5]

The multi-party congresses, in both cases, invited representatives of active political parties to negotiate solutions to the conflict. Each representative had to commit to democratic principles of dialogue and reject any use of violent means to advance political objectives. These congresses were moving from one crisis to another and could not finalize agreements. Moreover, violent episodes accompanied the whole process. However, the major contribution of the multi-party talks was to involve the people in the peacemaking efforts.

The multi-party talks provided a political alternative to the violent struggle and stimulated public debate on critical issues. At the end of the day, the people began to replace the conventional wisdom that there is no solution to the conflict with the idea that the struggle is resolvable. This was the sign for the leadership to conclude agreements.

In both cases, Northern Ireland and South Africa, political leaders—who understood their serious limitations—created the multi-party talks as an instrument to involve the people in the struggle for change. In the Israeli–Palestinian situation, political elite diplomacy (such as the Oslo Accords) and unilateral initiatives (such as the Israeli disengagement from the Gaza strip in 2005) are still the dominant peacemaking channels.

In the Israeli–Palestinian case, the public has not been involved in the peacemaking struggle, prepared to cope with crises during the peacemaking road, and so is not ready for a new social order. As a result, the moderate majority is desperate, and radicals, who have a clear agenda, shape public discourse. As a desperate choice, ordinary Israelis and Palestinians joined efforts to establish the Congress of the Israeli–Palestinian people.[6]

The Congress of the Israeli–Palestinian People

The Congress of the People includes a council of 120 Israelis and Palestinian—from different sectors and locations—who came to negotiate alternative solutions to the conflict. The goal was to reach negotiated peace agreements within half a year. The "official" negotiations took place once a month in a different location, and they were open for public participation. The members of the assembly used digital tools to negotiate online and prepare the formal meetings.

The negotiations focused on three alternative solutions to the conflict:

1   **Two State Solution**—Division of the land between the State of Israel and the State of Palestine in the West Bank and Gaza.
2   **Federation**—Israel and the West Bank turn into a federal democratic country: A central government becomes the primary political unit, uniting partially self-governing territorial autonomies. Gaza will be an independent state.
3   **Confederation: Two Cooperative States**—Two independent states—Israel and Palestine—operating under a Confederate roof, which enables them to establish joint peacebuilding and peacekeeping mechanisms.

The climax of the project was supposed to be the first convention of the Congress of the People on April 3 on Rothschild Boulevard, Tel Aviv. The plan was to present the three alternative agreements to hundreds of Israelis and Palestinians, engage them in a critical discussion and invite them to vote in ballot boxes. The organizers of the congress intended to publish the results and send them, with recommendations, to the leaderships. The Corona crisis has stopped the whole process and forced the organizers to change plans.

The Corona crisis has created a new reality—only small meetings have been allowed to take place. The Israeli and Palestinian founding members of the congress ("the Cabinet") decided to hold negotiating assemblies in a different format.

The Corona Crisis

On March 6, the Cabinet of the Congress—20 Israelis and Palestinians—met in the Arab-Hebrew Theater in Jaffa (Israel) to examine alternative ways to cope with the Corona crisis. They discussed and negotiated the question: How can we cope with the Corona Crisis in the frameworks of the three alternative solutions to the Israeli–Palestinian conflict?[7]

Each of the groups provided a different answer:

Two State Solution—Each country, Israel and Palestine, has to cope with the crisis by itself. Cooperation is a matter of goodwill and agreements between two countries.

Federation—The citizens of the Federation (Israelis and Palestinians) have a joint problem. The central government should coordinate efforts with all of the autonomous regions in order to cope, effectively and efficiently, with the coronavirus pandemic.

Two Cooperative States (Confederation)—The Corona crisis is a trigger to establish joint economic mechanisms and health committees to enable each country (Israel and Palestine) to cope with the crisis in the most effective way.

The three different answers clarify important aspects of each of the alternative solutions and its practical implications. Let me demonstrate by presenting central arguments.

The supporters of **Two State Solution**—separation between Israelis and Palestinians—believe that Israeli hospitals should serve Israeli citizens. They are afraid that in a case of a widespread epidemic, any other solution will lead to Israeli hospitals being full of infected Palestinians. This situation can have negative effects on the quality of medical care for Israelis. The supporters of **one state solution** argue that the Corona virus is not subject to human borders. It demonstrates again that separation between Israelis and Palestinians in the West Bank is impossible. A central government, which coordinates efforts with autonomous regions, is a just system for Israelis and Palestinians that can prevent the wide spread of the virus better than any other system. **The confederal solution** (two cooperative states) is an intermediate position. It includes separation (two national states) with joint mechanisms to build and keep peace and order. The Corona crisis is a precious opportunity to build the necessary cooperative peacebuilding and peacekeeping instruments. These tools intend to help prevent a situation where the new state of Palestine collapses similar to Gaza in 2007[8] and other new developing states during the 1960s (see Huntington 2006).

A detailed debate between the three camps (Two States, Federation, and Confederation)—which is beyond the scope of this paper—enables one to examine the advantages and disadvantages of each solution. Moreover, it demonstrates that "there is no free lunch." Each solution has costs and risks that must be weighed before making a decision. This is the main message of the Congress of the People to the Israeli and Palestinian leaderships and people.

The Question of Responsibility

The Corona crisis has the potential to change the world. At this stage, it is difficult to fully estimate and understand its impact. However, it is quite reasonable to assume that major crises, such as the Israeli–Palestinian conflict, are not going to vanish or to be resolved spontaneously. The Corona crisis can be an opportunity to examine different solutions to the conflict and better understand the advantages and disadvantages of each one of them.

The Congress of the People demonstrates that a Public Negotiating Congress—which includes people from all walks of life—can be a market place of ideas. It enables the participants to examine old problems from different viewpoints, to analyse the cost of each negotiated solution and brings new ideas to the discussion table.

It is the moral obligation of leaderships and ordinary people to be engaged in such an endeavour. However, moral obligation is not a strategy. There is an urgent need to develop a strategy that can bring the spirit of the small negotiating assembly in the Hebrew-Arab Theater—which uses the Corona crisis as a case study to examine advantages and disadvantages of alternative solutions to the Israeli–Palestinian conflict—to every house in Israel and the Palestinian society. This is an urgent call for leadership and citizen responsibility, which could be complementary to Ananta Kumar Giri's calling for Global Responsibility.

On the Decolonial Turn in Global Responsibility

Sabelo J. Ndlovu-Gatsheni, University of Bayreuth, Germany

Ananta Kumar Giri's *The Calling of Global Responsibility: New Initiatives in Justice, Dialogues and Planetary Realizations* is an epitome of responsible scholarship in the first instance. The responsibility goes beyond merely seeking to know the world and is escalated to the higher level of making a contributing towards changing the world. This type of responsible scholarship is very urgent and necessary today because as humanity we are experiencing a crisis of coexistence. In reading Giri's book, I saw a humanist in action—making a planetary call for global responsibility. The outbreak of Corona virus and the COVID-19 pandemic makes Giri's call even more important and relevant. The Coronavirus hit at the very centre of the planetary human entanglement, waking us up from the slumbers of racism, narrow nationalism, nativism, and xenophobia that is represented by Trumpism at a world scale. Giri writes:

> It is by now incontrovertible that we live in a more globally interconnected but fractured, contentious and fragmented world but what should be the character and direction of this evolving globality? Should globalization mean only economic globalization, and even corporate globalization, or should it mean humanization and planetary realizations?

These are very relevant questions which also drive my own intellectual work. Therefore, my responses to Giri's questions are two-fold. The first response is to name the problem we are facing correctly. The key problem is what James Blaut correctly described as "the colonizer's model of the world." What was delivered was colonial modernity predicated on the paradigm of difference (racism, patriarchy, and sexism), will to power, paradigm of war (homo polemos/warrior tradition), paradigm of discovery, and survival of the fittest. The consequences of these logics were genocides, enslavement, feminization, colonialism, and capitalism. If we go by these logics not the rhetoric of Euro-modernity (salvation, progress, civilization, modernization, development, and emancipation), it was inevitable that the humanity and the world would eventually plunge into the present crisis of Anthropocentrism. The warning was given long ago by the father of decolonization Aime Cesaire in *Discourse on Colonialism* (1955) in the following convincing words:

DOI: 10.4324/9780429347481-14

A civilization that proves incapable of solving the problems it creates is a decadent civilization.

A civilization that chooses to close its eyes to its most crucial problems is a sick civilization.

A civilization that plays fast and loose with its principles is a dying civilization.[9]

Here we are facing a civilizational crisis which requires us to rise adequately in terms of global responsibility called for by Giri. Thus, my second response to Giri's questions is to offer the "Ten-Ds" of decolonial turn namely:

- Deimperialization,
- Desecularization,
- Depatriachization,
- Deracialization,
- Debourgeoisement,
- Decorporatization,
- Democratization,
- Decanonization,
- Deborderization,
- Deanthropecentrism.[10]

These are key tenets to underpin global responsibility. The "Ten-Ds" of the decolonial turn enable envisioning of a decolonial model of the world as opposed to the colonizer's model of the world. They enable a decentring as a decolonial move which confronts the Eurocentrism which continues to sustain Europe and North America as the centre of the modern world within a context of increased planetary human entanglements, which are undercutting and subverting the colonizer's model of the world with its manufactured continental and national boundaries borne out of the paradigm of difference and capitalist logics. It is well expressed by the leading African intellectual Ngugi Wa Thiong'o in *Moving the Centre: The Struggle for Cultural Freedom* where he articulated three imperatives in the process of moving the centre: (1) moving the centre from its assumed location in the West to a multiplicity of spheres in all cultures, (2) moving the centre within all nations from the dominant social stratum of a Europeanized male bourgeois minority to the majority of peasants, workers and women, and (3) moving the centre from restrictive walls of nationalism, class, race and gender.[11] Decentring aims at making it possible for a pluriverse to emerge, a world in which many worlds coexist.[12] Ngugi wa Thiong'o speaks of "globalectics" and explains it this way:

> Globalectics is derived from the shape of the globe. On its surface, there is no one centre; any point is equally a centre. As for the internal centre of the globe, all points on the surface are equidistant to it—like spokes of a bicycle wheel that meet the hub. Globalectics combines the global and

the dialectical to describe a mutually affecting dialogue, or multi-logue, in the phenomena of nature and nurture in a global space that's rapidly transcending that of artificially bounded, as nation and region. The global is that which humans in spaceships or on the international space station see: the dialectical is the internal dynamics that they do not see. Globalectics embraces wholeness, interconnectedness, equality of potentiality of parts, tension, and motion. It is a way of thinking and relating to the world, particularly in the era of globalism and globalisation.[13]

The point here is that there is no global responsibility without what a new consciousness of the new era we live in. This is why such a consciousness would not emerge without first of all a clear understanding of the problematic colonizer's model of the world as the generator of most of the problems of today (see Fanon 2004).

Towards Abandonment of the Colonizer's Model of the World

Aime Cesaire's student Frantz Fanon became the most eloquent theorist of decolonization and the most sophisticated critic of the colonizer's model of the world. Fanon was concerned about what he termed the "European game" as a central leitmotif of the colonizer's model of the world. Fanon called for abandonment of the "European game" in these profound words:

> When I search for Man in the technique and style of Europe, I see only succession of negations of man and an avalanche of murders. The human condition, plans for mankind and collaboration between men in those tasks which increase the sum total of humanity are now problems, which demand true inventions. Let us decide not to imitate Europe; let us combine our muscles and our brains in a new direction. Let us try to create the whole man, whom Europe has been incapable of bringing to triumphant birth. Two centuries ago, a former European colony decided to catch up with Europe. It succeeded so well that the United States of America became a monster, in which the taints, the sickness and inhumanity of Europe have grown to appalling dimensions. Comrades, have we not other work to do than to create a third Europe?[14]

However, James Blaut is credited with conceiving the concept of the colonizer's model of the world which Fanon was criticizing.[15] At its centre is a Manichean structure which releases binaries and politics of alterity. This problematic binary structure was reflected in the *res cogitas*, and the earlier Manichean scepticism of *ego conquiro*. As an epistemological structural binary for thought, it can be traced back to the birth of western philosophy in Socratic rhetoric, and the Platonic world of ideal forms. Western metaphysics in asking "What is" already generated by implication "what isn't."[16] These logics enabled and justified brutal systems of slavery, imperialism, colonialism, and recently, global capitalism to perpetuate the myth of its superiority over the modern world. What emerged from this western "endarkenment" (peddled as "enlightenment") was the presentation and centring of western local histories, as global designs.[17] This brutal centring of western local histories as "universal," meant that all other histories, knowledges and conceptions of being/ontologies/subjectivities that fell outside of western logic of rationality, were deemed barbaric, parochial and provincial—and thus non-existent.[18] As Blaut points out, western modernity was forged on the myth that "the world has a permanent geographical centre and a permanent periphery: an Inside and an Outside. Inside leads, Outside lags. Inside innovates, Outside imitates."[19] That being said—we know it is the case, acting to change it shouldn't be contingent on accurately pinpointing its emergence historically.

The decentring of Eurocentric global designs is predicated on the colonial matrices that include, but are not limited to, colonialities of power, knowledge, being, nature, and space.[20] Coloniality of knowledge, predicated on the invasion of the mental universe of the rest of the world produced coloniality of power, characterized by the long-standing asymmetrical global power relations that have fragmented the world into two zones; the centre and the periphery (with the former being the zone of being, and the latter, the zone of none-being). Those country's societies occupying the centre enjoy the privilege of dictating to and dominating the periphery in all fields of engagement from the geopolitical, right through to economic designs. Being at the nucleus of the world, the developed western countries have also continued to impose global capitalism as the only legitimate ideologic-economic design to be followed by all and sundry.

Informing these asymmetrical global power relations is the *coloniality of knowledge* predicated on the myth of the West as the only legitimate producers of real knowledge. This deliberate centring of Eurocentric knowledge was done through the use of brute force that systematically imposed slavery, imperialism, colonialism, global capitalism, and the long-standing effects of these oppressive systems of subjugation, referred to as coloniality.[21] As the coloniality of knowledge is centred, it pushes, ignores, and/or kills all other knowledges that are produced in its periphery. Geographically and temporally—as in the denial of coevalness perpetrated through the universalized regime of western history that conceived of the periphery as chronologically behind through a logic of civilizational developmentalism/presentism. This means, for example, that the knowledges about landscape architecture, medicine, law, science, and physics that do not fall within the Western conception of Knowledge are rendered useless. What emerges from these epistemicides (i.e. systematic suffocation and decimation of othered knowledges) therefore is the robbing of the world of different ways of knowing, sensing, viewing, experiencing, and solving issues. It is this myopia of the globalized Eurocentric paradigm that constitutes the ongoing administrative inability to effectively think outside the colonial social systems generating climate change/Anthropocene.

The *coloniality of being*, on the other hand, is characterized by western global designs predicated on the ideology of "impossibility of co-presence" between people of different races, ethnicities, and sexual preferences.[22] Defined by radical difference, coloniality of being uses racism and racial hierarchization as the primary principle for geopolitical and spatial ordering of the world.[23] Flowing from this abyssal thinking[24] is the unleashing of both structural and systemic violence against "non-western" human bodies that find themselves having to contend with "hellish existence" as they languish in varied conditions of socio-spatial and economic marginalization and related markers of precarity and alterity that naturalize the "non-ethics of war."[25]

Since the idea of western civilization and progress has for over the last 500 years been Anthropocene-oriented; it has resulted in the plundering, looting, and primitive extraction of minerals, water, oils, and other natural bodies referred to in capitalist lexicon as "raw materials" and/or "resources." This treatment of non-human bodies as resources for furthering capitalist interests and profits has contributed to the destabilization of the balance between the human and the non-human world, and the destruction of the natural environment. This primitive extractivism has been contingent on a despiritualization and rationalization of the land, through the Eurocentric epistemological regime that sees the world as an object it seeks to know or instrumentalize.

For decolonial thinkers, this centring of the human as the master of the universe stems from the logic of *coloniality of nature*. Primitive extractivism at a global scale has resulted not only in the destruction of the natural environment but also in the perpetuation of wars, as capitalist states, companies and firms pit communities against each other, all in the name of extracting blood minerals and fossil fuels.[26] While these logics of coloniality are often discussed separately, they form part of a complex power matrix that is critical for the autopoietic functioning of the global coloniality. Addressing the ecologically, degradation constituting the Anthropocene then necessarily requires decolonizing.

The Decolonizers Model of the World—The Significance of the Decolonial Turn for Global Responsibility

Having sketched out the matrix of colonialities, I propose at least ten decolonial logics (Ten-D's) that can contribute to the re-designing of the world. These reconstituting logics are Decentring, Deimperialization, Desecularization, Depatriachization, Deracialization, Debourgeoisement, Decorporatization, Democratization, Decanonization, and Deborderization.[27]

Deanthropocentrism: this involves the removal of the human as the centre and master of all species in the planet, and the re-imagining of a "posthuman" world defined by a mutual and reciprocal relationship between the human and the non-human bodies. As the human/non-human line is dismantled, the logics of coloniality of nature, and power will dissipate, making way for the imagining of mutual economic relations between mankind and the natural environment (and the reinstating of mankind to nature of which s/he forms but a part).

Deimperialization of global designs and power relations: this refers to the decolonization of asymmetrical power relations through the decentring Eurocentric modernity. Part of that involves the provincialization of Europe and North America and as part of designing global power relations and the geopolitical space at large.

Desecularization of spiritualities of the world: this refers to the re-centring of all spiritualities which enabled humans and nature to coexist harmoniously with a world exposed to worldwide ecological crises. Philipe Descola and Bruno Latour have shown that specific conceptions of Nature are actually belief systems foundational to western epistemology.

Depatriachization of heteronormative colonial gender designs: this involves dismantling patriarchy and toxic masculinity and the treatment of women as lesser people. Gender as a modern organizing principle that inferiorizes women and reduces them to the most exploited people in the world must come to an end.

Deracialization of everyday geopolitical and spatial designs: this refers to the dismantling of the racial, ethnic, and colour line at a global scale. It is premised on the recognition of common humanity and the total banishment of all forms of discrimination. From a spatial design perspective, this involves planning for racially inclusive communities, cities, and the world at large. Racism must cease to be used as the primary organizing principle of ordering space.

Debourgeoisement of everyday life: this refers to an urgent need to shift away from making bourgeois way of life a planetary template for all lives. As the dominance of a handful wealthy nation-states continues to impoverish the rest of the world through asymmetrical power relations; new modes of living that promote co-existence must be charted and promoted at a planetary scale.

Decorporatization: the logic of the rule of market coloniality leading to the commodification and commercialization of everything including life has weakened investment in our health systems and human security. We call for a shift from economies of profit to economies of care.

Democratization of the world: the present asymmetrical and pyramidal structure of global power with a sole superpower supported by accumulation of weapons of mass destruction is inimical to democracy.

Decanonization of epistemic designs: this refers to the decentring of Eurocentric knowledge that for centuries has canonized itself as universally applicable knowledge. Decanonization therefore is meant to bring into the fold all knowledges that were pushed to the margins by the colonial epistemic regimes. This also means embracing relevant knowledges from indigenous people's archives, towards the creation of pluriversal ecologies of knowledges.[28]

Deborderization for mutual global mobility: this refers to the shift away from narrow nationalism that has resulted in the hard borderization of the world as well as the resurgence of anti-immigration sentiments across the world. As the world has always been defined by migrations, it is non-sensical for nation-states to systematically prohibit the mobility of people, while on the other hand deliberately promoting the movement of capital across borders.

Conclusion: Towards Global Responsibility

What is offered here is a decolonial ethical socio-political framework necessary for a postcolonial and postracial world characterized by heterogeneous intersubjective relations underpinned by planetary human entanglements. Such (ir)responsibilities as racism, sexism, patriarchy, and anti-blackness have no space in such as pluriverse. The "Ten-Ds" of the decolonial turn enable a new designing of another world where co-existence between all species (human and non-human) is possible and therefore will involve unlearning some disciplinary approaches (about design) in order to re-learn new ways of seeing, viewing, experiencing, and imagining inclusive utopian registers. After all, another world is possible where justice and responsibility reigns![29]

Some Key Concepts in Moral Debates

Postface to the Calling of Global Responsibility

Rico Sneller, Leiden University, The Netherlands

Part of the challenge any call for global responsibility is faced with is a reflection on its condition of possibility and corresponding premises. Such a reflection should be modest, and not repress but rather stimulate responsibility awareness. In what follows, I will introduce four indispensable moral notions and relate them to Giri's account of global responsibility: "retrospection," "recollection," "memory," and "forgiveness." For these notions, I am partly drawing on my book *Perspectives on Synchronicity, Inspiration, and the Soul* (see Sneller 2020) However, when giving them a central place in this postface, I must equally admit my debt to the heritage of Indian philosophy, which has always had a keen eye for *consciousness*. Concepts never exist in themselves as self-contained entities; they always entertain a relation with consciousness.

To the extent that consciousness must necessarily be taken as stratified (it can always expand), notions like "ethics" and "morality" cannot remain unaffected themselves by this stratification. Unfortunately, our standard moral discourse today—its concepts, distinctions, and premises—is strongly characterized by a flat (conception of) consciousness. The challenges we are facing today, however, might be particularly addressing other levels of consciousness. For example, the problem of "global inequality"—which has been discussed extensively in Giri's book—would be handled inadequately if it were only seen as a subject matter that can be clearly defined in terms of objectifiable *external* phenomena—as if global inequality were an "object" for a "subject." Instead, "global inequality" (to stick to this example) should first and foremost be framed as a form of growing self-*alienation* in those defining and addressing it before being addressed by it.[30]

In light of the prevailing and yet misleading flattened conception of consciousness, we had better interpret "ethics" in terms of *implication*. An issue is ethical or moral if, and only if, a stratified consciousness implies it. The verb "to imply" literally means "to include" and "to enfold." In my view, the ethical is that which results from the unfolding of an inclusive, condensed consciousness. The *incentive* which I believe is inherent to anything ethical—both in its distorted, flat conceptions and the derivative conception which I am proposing here—unfolds in proportion to addressing of consciousness as multi-layered. Truly *ethical* questions are fundamentally different from,

DOI: 10.4324/9780429347481-15

for example, "prisoner's dilemmas" or "trolley problems" used by university teachers in class to introduce their students to the field of ethics. These questions cannot fail to affect subjects—and unravel them *as* subjects. I would suggest that the reverse side of this unravelling is that veridic ethical questions are indicative of an ongoing maturing process. True, the logic of maturing processes often seems completely opaque, as they usually entail human suffering.

When discussing ethical questions, I am inclined to exploit the ambiguity of the word "question." An ethical question is not merely analogous to a question which demands a symmetrical answer, such as "What is the capital of Maharashtra?" An ethical question is rather a *quaestio*, a topic, or an issue to be considered. This consideration could be painful; the Latin *in quaestionem dare* means "to extradite for torture." The intended result of considering a *quaestio* is not just an answer but gain or profit (*quaestus* means "gain," "profit," and "advantage"). Dealing with ethical issues entails suffering and gain, in other words, "maturing." If there is neither suffering nor gain involved, the so-called "ethical questions" at hand are either *mal*-treated or disguised economic questions.[31] In Giri's account, we have seen how relevant it is to stay alert when economic interests are at stake.

I will now turn to the four moral notions which I think are indispensable when discussing global responsibility.

Retrospection

Let me quote once more the words of the Indian thinker Maulana Wahiduddin Khan, also quoted in the preceding by Giri:

> What is the point in lamenting about our [Islamic] achievements in the past? We should go ahead with our efforts to advance and progress in the present. We should go ahead with modern science and technology. This is also the gift of Allah. Muslims should learn science and technology and be part of the modern world.

I interpret this statement not as a call for forgetfulness. On the contrary, Maulana calls for a conscious and more mature relation to one's past. I propose to use the term "retrospection" here. Retrospection, in my view, consists of seeing the past as present in the present. An intensified focus might bring out this past; not as an impediment but as a condition of possibility. It *might*, yet *need* not, bring out this past, as there will always be a threat of fixed ideas which force one to disentangle these before any progress can be made. I believe that so-called hostile views (e.g. of "Americans," "Jews," "Arabs," "Bengalis," the "working class," or "capitalists") are primarily based on fixed ideas, meaning that history has interposed—and continues to interpose—a variety of diaphragms which obscure reliable hindsight.

When I define retrospection as seeing the past as present in the present, I do not intend to identify past and history. What is called "history" consists of the development of the past towards an unexpected and unpredictable outcome. The Islamic past, for example, may hold in store treasures that can promote economic welfare in a way which is hitherto unseen. Obviously, the same applies to the Hindu past or the Jewish past.

Recollection

My second moral key term is "recollection." "Recollection" refers to re-digesting past memories. My claim is that the act of recollection can ultimately re-create, rather than just repeat, reality. The malleability of "reality" consists of the susceptibility of self and world to endlessly altering connections and modifications.

To illustrate my point, I will refer to the tragic events in Brazil related by Giri:

> The struggle for land and dignity and a fuller humanity in MST [*Movimento do Trabalhadoressem Terra*, that is, landless workers] has been accompanied by much violence and bloodshed. On April 17, 1996, 19 workers of MST were massacred while doing peaceful protest in Para. This massacre is being remembered by MST and Via Campesina as an International Day of Peasant Struggles. This preceded another massacre at Corumba in 1995. In 1997 alone confrontation between police and MST workers led to two dozen deaths.

Recollection, however, need not limit itself to redigesting past trauma. Instead, it could contribute to resituating the past in an enlarged consciousness and perception of self and other. The International Day of Peasant Struggles was introduced to "remember" the massacre. This commemoration is meant to be future-oriented, as we have seen; its concomitant "mysticism 'reduces the distance between the present and the future, helping us to anticipate the good things that are coming' (*Jornal Sem Terra* 102: 3)." Drawing on the possibility of an always widening consciousness, I would argue that the "recognition" implied by remembering can go as far as perceiving the past in a new light, if not even as acknowledging it. Recognition, then, will paradoxically be a confrontation with a deeper layer of the past which has never even been present, that is, never duly recognized in a present perception. The implication of this paradox is, first, that the past (here: the subsequent massacres in Brazil) may keep something in store the essence of which is still to be apprehended, and second, that this stored secrecy is somehow connected with the self (here: the self or shared consciousness of the landless Brazilian workers)—why else call it *recognition*?

Memory

The third moral key concept I would like to mention here is "memory." Trustworthy remembering and reliable memory, I would argue, are characterized by spontaneity. Memories emerging without the intervention of will, let alone the interposition of fixed ideas, are more reliable than repetitive obsessions. Again, this reliability does indeed not imply an unalterable past susceptible to seamless reproduction by the mind. Instead, it implies a past susceptible to recognition *beyond* mere reproduction. If, as is my claim here, consciousness cannot be delimited, the faculty of recognition originating in it becomes abysmal and a source of renewal, or a renewed discovery of the past. The past still has not passed, let alone that it has passed away; it is always yet to come. I believe that any act of commemoration, vital though it be, should not be obsessed with "facts" or a "fact world": for example, a supposedly "irreversible" slaughtering of landless labourers in 1995, 1996, or 1997; this could all too easily lead to revengefulness, the development and nourishing of fixed ideas, and sometimes even to nihilistic despair. An ideal "commemoration" would be disorganized and spontaneous.

For an illustration, we can think of the Brazilian children's initiative described in the same context:

> In 1998, one State-wide meeting in Santa Catarinaopened with several children walking single file through the audience carrying the tools and fruits of working the land—a machete, a handful of beans, a large squash. These were all laid at the front on a large outline of Brazil, signifying the construction of a better nation through the practices and values of MST's new community.

Forgiveness

My fourth moral concept is "forgiveness." Giri already referred to it by claiming that "[w]e all appreciate in others the inner qualities of kindness, patience, tolerance, forgiveness and generosity," and that these qualities all root in the core value of compassion. Compassion and forgiveness, I would insist, are intrinsically connected and mutually implicative. There is no compassion without forgiveness and vice versa.

Just as my three other key terms, "forgiveness" combines morality and consciousness in a single concept. Forgiveness is more than mere pardoning, especially when the latter is taken in the (non-moral) sense implied by the exclamation "I beg you pardon!" In forgiveness, something is given or revealed *indirectly*, something else being taken away (e.g. the massacre of innocent victims). What is given or revealed can only be noticed (if at all) *upon* the removal of something else. This "something else," I argue, is *time*. Or at least, time is taken away *along* with what is immediately taken away (e.g. something past). What is usually called "time" would at that point—instead of an incessant continuity of instants—*rather* represent a discontinuity, susceptible to hosting another, far more pressing "temporality."

Forgiveness recreates the past by fathoming its endless malleability. To use the beautiful words of the completely forgotten German philosopher Hermann Friedmann (1873–1957): culture represents "the ethos of the human eye that is living-towards a light which becomes form in heaven."[32] In my interpretation, this means that culture consists of a double promise: both the promise of future illumination and the promise of a future transformation of this light into concrete form. Once the prospective illumination has received a heavenly form, it will be adequately informed. The corresponding ethos (forgiveness) will be an ethos which anticipates this future transformation and is justified by it. This ethos entails restless endeavours into obscured otherness, always recommencing and always bringing to light what is new.

Obviously, it would be utterly *immoral* to *impose* forgiveness on victims of cruelty. Who does not remember the outcry of Elie Wiesel at a commemoration of the liberation of the Auschwitz concentration camp: "O Lord, do not forgive the murderers of our children!" Simon Wiesenthal, another Jewish survivor of the death camps, reminisces that he once refused forgiveness to a dying SS officer who had asked him for it; he felt unable to forgive *on behalf* of those who had already been murdered. While Wiesenthal has subsequently been blamed for withholding forgiveness, and while he himself kept struggling with the event throughout the rest of his life, I think that some cruelties are too big to be forgiven, at least by a human being. However, *if* and *when* forgiveness is viable, it can repair damage in a way nothing else can.

Conclusion

In this postface, I have tried to reflect on Giri's account of global responsibility by highlighting a set of moral concepts which I think are indispensable: retrospection, recollection, memory, and forgiveness. These concepts have in common that they not only represent a key value for moral life but also indicate the share which consciousness has in them. "Thin" moral language consists of bare concepts erroneously taken as tools to articulate moral life; to make this "thin" language "thick," it is requisite to account for the way consciousness is involved in it.[33] By adducing some examples from moral situations mentioned in Giri's text, I have tried to show that the author is implicitly aware of this. Insofar as the traditions of Indian thinking tend to pay tribute to dimensions of consciousness, I am convinced that they enrich narrow-minded, over-conceptualized forms of reflection which largely dominate moral debates today, especially when conducted in academia.

Ethics of Globality and the Burden of Responsibility

Mahmoud Masaeli, Alternative Perspectives and Global Concerns, Ottawa, Canada

I was glad when I got a copy of *The Calling of Global Responsibility: New Initiatives in Justice, Dialogues, and Planetary Realizations* authored by Ananta Kumar Giri. The title itself is inspiriting enough for anyone who feels responsible for human suffering especially structural injustice in an age of unbridled capitalism.

Two thought-provoking themes in this volume provoke deep contemplation on the nature and the condition of the current lifeworld. The first is the condition of globality, or as Ronald Robertson puts it, the condition of glocality. This means that the world is experiencing a process of fundamental transformations that is eroding the normative importance of all territorial-based concepts, perspectives, and approaches. Indeed, in the condition of global interconnectedness and networking, not only the geography of wealth, happiness, and fairness in social transactions are being changed but also the conception of proximity is gradually being connoted a new meaning. The COVID-19 pandemic is a good illustration of the changing nature of proximity. Although the virus began in China, today most of the western countries, among them the USA, are in the most severely affected condition. This example reflects how the condition of globality is weaving the world together. The example also proves that the condition of globality is like a teleological process which is weaving the life all around the world without a chance to escape. Other global issues such as ecological issues, environmental degradation, trafficking of drugs and human organ, money laundering, and the xenophobia of different sort embodied in the right-wing groups, are turning into global threats engendering the life of all in the globe. Yet global consciousness conducive to the emergence of globalization of masses, grassroots movements, and the organizations of global civil society cannot be ignored. It is obvious that for the first time in modern history, people, at least theoretically, are able to express themselves against the patterns of human suffering. Global consciousness is rising to the highest level of global institutions and structures. In the world of agony, there is a chance to raise up and try for humanity. In this perspective, the condition of globality signals the emergence of a networking global community in which the oppressed people can stand up and strive for recognition of their differences.

DOI: 10.4324/9780429347481-16

In this condition, territorial-based views of the world, in whatever way one imagines, are losing their validity necessarily because the geographical boundaries are becoming permeable. Therefore, local manifestations of selfhood and nationhood, value systems and cultural outlooks, living conditions, lifestyle, and tasks and responsibilities are being affected by global waves of transformation. At the same time, globality cannot escape itself from locality, since local forces are able to express themselves in global settings. This state of the mélange of globality and locality has provided new opportunities to plan, act, and promote either positive ideas for humanity or negative intentions against the community of mankind. This state of mélange is bewildering as to where we are going. Are we going towards a condition of rampant individualism, unbridled capitalism, more social exclusion, and marginalization or the world is moving towards a condition of responsibility for the excluded others, curing inequality and injustice, and the formation of a "global we"; a cosmopolis? In terms of its appearance, globality signals hope and promise for ending the historical patterns of suffering; capital flows, cultural flows, internet and telecommunication, cultural flows, globalization of human rights norms and the significance of democratization, and global consciousness. However, in terms of its content, globality reflects the hegemonic global capitalism and irresponsibility of transnational corporations, global patterns of exclusion, and cultural diffusion of the rampant individualism. This condition of uncertainty on the effects of globality for our common humanity is associated with ebbs and flows; paradoxes, risks, and threats, on the one hand, and the opportunity to build dams on the river to generate power, on the other.

The condition of globality and its paradoxes inspire the second cardinal idea. Who is responsible for the accelerated human suffering; inequality and exclusion, systemic risks, and fragile systems? While the benefits of the condition of globality are basically presented by predominant global ruling classes, who must resonate with the voice of the excluded, marginalized, and suppressed? What is the nature of this responsibility; adjusting the world, deconstruction of the world, or striving for humanization? In what way is the responsibility directed? Is the direction symbolized by the invisible hand capitalism inspiring and effective? Who burdens responsibility? In the mainstream social sciences and humanity as well as management and leadership, responsibility is identified with the appealing conception of "help." The help, which is becoming a culture, and industry, is seemingly a great idea because the receivers of the help will be assisted to reduce the severity of their hardship, while the helpers are also able to get a sort of inner satisfaction and moral perfection. In all cultures, the conception of help entails moral responsibility. That is why many have been able to justify their unjust actions behind the culture of responsibility for others through helping them. Truman, in his inaugural speech, the fourth point, on January 20, 1949, manifested this responsibility. For him, who became a pioneer in the post-WWII culture of help and responsibility, the face of the world was portrayed in the responsibility of the advanced societies to help under-developed nations. Regardless of ignoring the systemic causes of unequal development in the world caused by the complex patterns of colonialism and imperialism, in his appeal to celebrate global responsibility, the exploit-ative nature of the international system was, perhaps, wittingly overlooked. On that day, two billion people were labelled underdeveloped, people who are intrinsically incapable of advancing their own living conditions, that is, needy people who must be directed to a happy life by the west. Soon, such a culture of moral responsibility, which turned to an atmosphere of the ideo-logical rivalry of the West against the Communist block during the Cold War, led to the emergence of the discourse of responsibility; we are respon-sible to help others reduce their suffering. Furthermore, the responsibility to help became a profitable business or the deceptive means for promoting prestige, or even, alliance-making in the international arena.

Nevertheless, the historical patterns and structure of dependency of the disadvantaged nations to the former colonialists, today's crusaders against underdevelopment, were neglected. The discourse of moral responsibility crafted in the mind of people of non-advanced nations a submission to the saviour West led by the USA. It is only a moral responsibility to reduce the severity of underdevelopment and poverty, and not a duty for justice; just international order. As a sacred discourse, and through colonizing the mind of people, this self-acclaimed moral responsibility undermined the foun-dation of duty for justice. The old colonial patterns of domination were replaced by the invisible impression of hegemony over the mind of people. As a discourse, the responsibility to help disregards the systemic causes of injustice in a world that is been structured around the exploitative patterns of suppression and exclusion. Hence, the fundamental question yet persists is "what is the meaning of responsibility"? Perhaps better to raise two more fundamental questions: (1) Is it only a responsibility to help or a duty for justice? and (2) In either case, who must take the responsibility? The big questions are getting stimulating in the condition of globality. In this condi-tion, the raising global consciousness, one the one hand, and the possibility of being proactive for justice, on the other hand, instil a new spirit in the conception of responsibility and its bearers. These are not the states that are the only players in the world of inequality, irresponsibility, and the condition of agony. Nor are the global donor organizations and financial organiza-tions that can determine the rules of the game. The condition of globality empowers people, who are able to regroup themselves in global settings and create pressures for the fundamental changes in the structure of the world. The global movement of people, the revolt of masses, for justice entails a new form of responsibility. It is a responsibility to stop the systemic causes of human suffering, hence cannot be reduced to palliative solutions granted under the banner of morality to help. Imperative in this responsibility for justice is the means of achievement of the goal. Although the world is striv-ing for justice, the responsibility is not oriented towards violent behaviours and destructive solutions. The globality from below is wise enough to get into constructive means of changing the world. This is a point that has been raised in chapter three of Giri's book. To be appealing, responsibility for justice must be extended dialogically. Not only dialogue characterized the identity of people, the dialogical selves, but it also enhances their intellectuality and spirit towards a higher social level dialogue about the purpose of life. Dialogue empowers people to express themselves in social life. However, and more importantly, the dialogue is the constructive means of achieving justice. Dialogue improves our own sense of humanity, instructs informative social roles and responsibilities, and bridges the gap between perspectives over the meaning of life. In this sense, the dialogue turns to an internal–external mode of dealing with the issue of life in the condition of globality among them inequality and injustice. People, organized in the organizations of civil society, play the role of insiders suffering from injustice, and the states, and international players, as outsiders can come into a mutual understanding of the concerns, paradoxes, and threats. This sophisticated dialogue also breads responsibility to exchange ideas, reduce the costs of interactions in achieving the goals, and enhances the benefits for all.

The book by Giri has the potential to raise the fundamental questions about the conception of responsibility for injustice in the condition of globality as well as dialogical approaches for justice. These questions cannot be answered easily. Nor is a straightforward answer to the questions. But, for sure, the book is provoking deep thinking about globality and responsivity for global concerns.

Humanity in Face of an Uncertain Future: Are Gaia and PachaMama Going to Guide Us?

Responsibility Towards All Living Beings and Towards Nature Will Be the Key to a New Stage in Human Civilization

Beatriz Bissio, Federal University of Rio de Janeiro, Brazil

Ananta Kumar Giri's book invites us to reflect deeply on new paradigms, new utopias, from a central concept, **Responsibility**. These are important, necessary, and opportune reflections, because we are living an historical stage that challenges us to a critical evaluation of the reasons that led humanity to the critical situation in which it finds itself.

Giri defines responsibility as a key concept for the 21st century. I couldn't agree with him more! This statement forces us to think what responsibilities have been left aside, by whom and in relation to whom or what. And at the same time it calls us to think about our responsibilities as human beings who believe that "another world is possible," remembering the social movements' slogan raised at the World Social Forum.

The scientific evidence shows that we are close to the point of no return in relation to the destruction of our natural habitat (ten years, only, according to some scientific sources!), but it is also known that the application of today's human knowledge and technology would allow us to change this scenario. So, who is responsible for failure to act in this field?

According to the World Health Organization (WHO), there are currently more than 820 million hungry people globally (approximately 11% of the world population). "This underscores the immense challenge of achieving the Sustainable Development Goal of Zero Hunger by 2030," says the 2019 edition of "The State of Food Security and Nutrition in the World" annual report.

Whose responsibility is the failure in this field?

According to the same report, most people in the world live in poverty. Two-thirds of the world population live on less than 10 USD per day. And every tenth person lives on less than 1.90 USD.[34] The world's 2,153 billionaires have more wealth than the 4.6 billion people who make up 60% of the planet's population. "Getting the richest 1% to pay just 0.5% extra tax on their wealth over the next 10 years would equal the investment needed

DOI: 10.4324/9780429347481-17

to create 117 million jobs in sectors such as elderly and childcare, education and health," according to the same report.[35]

Who is responsible for the lack of action on this issue?

The answers to these questions have been long known. But the adoption of the necessary measures has never been given priority. Powerful controlling vested interests ignored these inconveniences. But the pandemic that has afflicted global humanity since the beginning of 2020 has forced a rethink of our old problems that from now on acquire urgent and obvious relevance. It has also unveiled the importance of finger pointing those responsible and defining initiatives to tackle the challenges. This poses practical and theoretical problems, as many certainties are crumbling like dust!

This new reality reminded me of the debate that heated up the 1960s and 1970s in some intellectual and student circles which I was part of, in the framework of the ideological struggle between those who defended the superiority of capitalism over socialism and those who supported the socialist or the communist ideals: what is more important in a society, freedom or equality?

Those who advocated the idea of the primacy freedom—as it was understood in the liberal democratic context—were basically defending the very foundations of the capitalist system. They were supporting the existence of a free, self-regulated market and of free initiative in economic terms, referring also to human liberty as a value. They were confronting and putting forward arguments to question almost exclusively the system in force in the Soviet Union, seeing it as a model in which "individual rights" were subject to collective logic. In contrast, those for whom equality was to be considered the most important element for life in society pointed to the fallacy of freedom in the capitalist system: by leaving the economy in the "invisible hands" of the market, capitalism could only offer, for most of society, an illusory freedom, because it only could be limited, a freedom which, according to a humorous metaphor quoted at the time, is reduced to the possibility of choosing between Coca-Cola and Pepsi-Cola. They stressed that a society in which there is no equality or in which the degree of inequality is evident can never be considered effectively free, since important segments of their population due to their fragility, live at the mercy of the arbitrariness of the most powerful.

Another debate raised in that period, but in fact as old as civilization itself, was what comes first, the society or the individual? Depending on the answer, very diverse behaviours are adopted to face the challenges.

Naturally, today we live in a very different context from the Cold War, in which these debates were held. However, the issues discussed are pertinent in the present, particularly in the light of what the COVID-19 pandemic has unveiled.

Our societies, in the Western world at least, when submitted to the neoliberal capitalist logic, are dysfunctional: not only are a majority of the human beings excluded from the minimum survival conditions but also irreparable damage is being caused to our common home, the planet Earth. The global virus provoked quarantine has forced the opening of the debate, at very different levels, on the absurdities of a system that does not allow banks to fail, but does tolerate the condemnation of millions of human lives to poverty and hunger, spending their existence without the slightest opportunity to really be part of society. And has shown how artificial the human-defined frontiers are, mostly an infamous colonial heritage, which did not respect cultures, traditions, history, and immaterial values.

With few exceptions, countries with fewer social inequalities have tackled the virus with correct measures, thus obtaining significant achievements. Another fact is also evident: the society used to prioritizing collective decision-making and action, has greater chances of quickly defining means to defend itself against contagion, even in the absence of guidance from the authorities. It appears clear that without an organized society, we, individuals, cannot survive. And it is obvious that societies in which social equality is not a value to be sought, will not survive either. This will be more and more clear in the future when, if the scientific forecast is correct new periodical pandemics are expected, and there is enough information to substantiate this!

In a moment of crisis, as it has been dramatically demonstrated in 2020, we all depend on each other, and no one is safe unless he or she has the help of others. Therefore, the importance of the responsibility to care! Only life in and as a society allows us to survive as individuals. Implicit in this truth is a concept that was already present in the thought of Confucius (551–471 A.C.) and Aristotle (384–322 A.C.): humans are social beings by nature. Besides, as rational beings, meaning enjoying free will, humans need to live in freedom. But historical experience has shown that our personal fulfilment depends on the fulfilment of everyone. Thus, freedom in its true essence can only be achieved in societies that permanently seek the ideal of equality.

That is why these reflections matter today. Not only in the short term, but due to the consequences they will have in the future.

As a former journalist and editor, I'm used to watching media coverage carefully, particularly in relation to national and international politics. In these months, the commercial media relationship of dependence on political and economic power has become more evident. In fact, this subject has been studied and denounced for decades. To understand this subject is vital and important to study the work developed by the International Commission for the Study of Communication Problems (appointed by UNESCO in 1976, chaired by Sean Mac Bride, and therefore also known as the MacBride Commission). The Commission's brilliant draft Report was presented during the 20th General Conference of UNESCO, held in Paris on November 28, 1978, and the final Report was published in 1980 under the title *Many Voices, One World* (Mac Bride Commission 2003). The document was the basis for the UNESCO's proclamation calling for the establishment of a New World Information and Communication Order (NWICO) and that was the main reason why the USA and the United Kingdom withdrew from the UN Agency.

As a result, UNESCO lost most of its revenue and was almost paralysed for decades. The recommendations of the Commission—calling for democratization of communication and strengthening of national media to avoid dependence on external sources, among many others—were never implemented. Since then, media concentration has only worsened and information flows have become even more unbalanced between the North and the Global South. With a few important exceptions the commercial media continues to be a powerful instrument of the production of meaning attached to ruling elites. This is particularly true in most of the peripheral world.

This reality is now in full evidence, with the coverage of the pandemic. The daily figures of the contagion—I am referring mostly to Europe, North America, and South America—indicate the highest incidence of the Coronavirus among black, indigenous, and marginalized people. The pandemic does not respect class, racial/gender definitions or even national borders. Therefore, the fact that the highest proportion of contagion is among blacks, indigenous people, and in the poorest urban areas indicates that the main reason lies in the socio-economic conditions and in the possibilities of access to infrastructure (water supply, basic sanitation, medical-hospital care). The lack of such infrastructure is related to historical deprivations of rights of the poorest people and, more recently, to the neo-liberal politics imposed on most Western countries. As a result, we have today a deeper social gap between the richest and the poorest. Should not these subjects be object of profound discussion by the media? Isn't that a prior responsibility?

It is necessary, thinking on responsibility, to debate alternatives to the imposed model and to define urgent policies to overcome the terrible social inequalities within and between countries; it is imperative to denounce the predatory practices that are destroying the environment and are exploiting and discriminating hundreds of millions of human beings. And the discussion of the need to judge those responsible for all these blunders (or should we, without euphemisms, call them crimes?) is unavoidable.

But these issues cannot be part of the mainstream media agenda for open and free debate. Most of them are hands-tied. They are either committed to that model or they are part of it. As a consequence, addressing the challenges of the Coronavirus pandemic the media only open the discussion of the most urgent short-term problems. They can't allow the socio-economic development model, the capitalist neo-liberal model, and its consequences to be put in check.

Today, the big bankers, the CEOs of the mega-holdings, the members of the 1% that exploits 99% of Humanity, and most of the oligopolistic media are facing a challenge: a small virus managed to reveal that all the certainties of their speeches were no more than a house of cards. Of course one cannot expect an act of repentance or "mea-culpa" from either of them.

If they do not encourage this debate, the society can. Responsibility!

Yet societies have been looking for alternatives for a long time. And they have formulated a mature diagnosis and proposals. There is no longer any doubt that the solutions to the planetary crisis that Humanity is facing will come from below. Today more than ever it is clear who is truly important in society. The bankers? The speculators? The landowners? No. On the contrary, in many countries—Brazil is perhaps one of the most representative—the elites have taken advantage of the anomalous circumstances created by the pandemic to withdraw formal workers' rights leaving them in a precariousness position. Furthermore, they are dismantling public policies, in favour of the private sector.

But, in the opposite direction, the pandemic has shown millions of doctors, nurses, health technicians, cleaning staff, standing by infected patients in hospitals, even at the risk of their own lives! Small rural producer and peasant farmers on the outskirts of the big urban areas, continuing to bring healthy and fresh products into the cities. The responsibility for the continuation of life remained on the shoulders of the small, the rejected, and the invisible. But the previously ignored role of those who did not use to be news became clear, evident, and the effect was quick. They were globally honoured with sympathetic and sonorous applause from all social segments.

The first initiatives to offer food and all kinds of support to the unprotected populations, to the thousands left without work by the pandemic, and to those who had lost loved ones also came from the spontaneous reaction of society. But supporting these spontaneous actions was the organizational experience of grassroots movements. This experience has a long history, from the many struggles in defence of life, land, unrecognized rights, against forced displacement. The universal reaction to these testing times have shown and are showing that popular movements emerge with more strength and effectiveness. In spite of their different affiliations and origins, grassroots movements have shown that it is possible to work together, organizing collectively on other bases, inverting priorities, decentralizing decisions, valuing who actually works and produces for the community. In Latin America numerous initiatives point in this direction: a new society, a dialogue of shared knowledges. These grassroots movements have been working on proposals for the social and economic transitions to post-extractivist models, for the abandonment of mono-culture, to ensure food sovereignty, for the option for an energy matrix that is no longer based on fossil fuels, a matrix organized in a decentralized way and outside the impositions of the market.

The same is happening in Africa and Asia, and an intercontinental or transcontinental dialogue among these movements is growing, particularly driven by the new generations, born with the idea of a global civilization, and by indigenous and black people. Significant detail is that, in most of these movements, there is a majority of women.

The indigenous people of Ecuador, Peru, Bolivia, define this utopia as the society of Good Living, in which there is harmony between human beings and the PachaMama, the Mother Earth. "Terricide" is the definition of the indigenous of Argentina for the current stage of society, a stage in which human society leads to the extreme destruction of natural habitat! This is why they fight for a radical paradigm shift: a new birth, a new learning. In this context, Latin-American indigenous communities advocate a new legal definition for the natural habitat, equating it to human beings: nature as a subject of rights! Humans have responsibility in relation to nature.

It is not by chance that the new paradigms are being proposed by the indigenous populations, the ones decimated by the genocide caused by colonialism and neo-colonialism. They advocate a pact for socio-environmental concerns, in defence of life in all its manifestations, through an alliance with all human beings committed to Justice in the most broad sense (historically, economically, socially, culturally, and so on). Nia Huaytalla, an indigenous feminist from Argentina, defined "climate justice" as "racial justice." For her,

> climate justice is the historic reparation that the privileged white people around the world owe to racialized people. After hundreds and hundreds of years of plundering our territories (the *Abya Yala*, name of the American territory given by the Kuna indigenous peoples) it is time to pay for what they have stolen.[36]
>
> Interesting and challenging remark!

Nia's statement reminded me of the sentence written by the American Pan-Africanist, sociologist, and historian W. E. B. Du Bois in the Forethought of his 1903 book *The Souls of Black Folk*: "The problem of the Twentieth Century is the problem of the colour line." Perhaps if he was alive and observed today's society, his sentiments and assessment would be the same as Nia's: The problem of this 21st century is still the colour line! Structural racism is everywhere. It is necessary to look for global responsibility to combat and overcome racism!

Symbolic attitudes have been taken all over the USA, Europe, and other parts of the world in relation to that responsibility: massive demonstrations, in spite of the Corona virus, and the destruction or removal of monuments and memorials of colonialists, slave masters, and figures connected with racism. The protests followed the May 2020 killing of George Floyd but developed into a wider denunciation of colonialism, racism, and imperialism. Claiming that his arrival in the Western hemisphere had marked the beginning of the genocide of Native American People, statues of Columbus were also removed in the USA. Reaction against racism and against all genocides is uniting in Brazil the different expressions of the black movement and the indigenous organizations in a process that shows the maturing of their understanding of how to analyse and face their age-old exploitation.

Are we foreseeing a new Era? Some analysts are seeing the present as the birth of something different; some of them are calling it a new Enlightenment.

A new Enlightenment? Even if we only used the metaphor with the aim of rescuing the good legacy of that period it does not fit to describe the transformation were are foreseeing. (Less noble developments of Enlightenment such as Eurocentrism, divorce between rationality and spirituality as well as white supremacy and all its consequences, can't be dissociated from it, at least not in today's world.)

The era that is being born seems different, because the main drivers of change are those called by Frantz Fanon as the *Wretched of the Earth*, those who suffered violence and dispossession but are aware of their strength and resilience. Their old wisdom is illuminating the birth of something new. But, as we are immersed in the process of change, the signs of the new are imprecise and the old resists being supplanted. That is why Antonio Gramsci's (*Prison Notebook* 1992) famous phrase fits in well to our present: "The crisis consists precisely in the fact that the old is dying and the new cannot be born; in this interregnum a great variety of morbid symptoms appear."

Even being difficult to define the future, it's possible to bet that it will mark the resurgence of the civilizations subjugated by colonialism for a painful period, in Africa, in Asia, and in Latin America. Civilizations which continue to cultivate communitarian values rather than the individualist focus of the global capitalism-based consumerism; communities which cultivate a deep relationship of respect and love for nature and have strict sense of responsibility for the future generations are under attack. But civilizations with a profound sense of communion with other living beings and with the natural world, with the Pacha Mama, Mother Earth provide us a perspective of lived spirituality. Science can conciliate with such spirituality! Let's here remember Gandhi's use of spirituality as a basis of social change.

The 18th- and 19th-century model of the world as a machine has led humans to the vocation to control nature; on the contrary, thinking of the Earth as Gaia, as Mother Earth, as a web of relationships between our natural habitat and life in all its manifestations leads to the responsibility to take care for it. It is a change of paradigm.

"*Cosmovision*[37] and spirituality are a simultaneous experience, acting at the same time, myth and history, death and resurrection," says the Maya philosophy. Only a human civilization with a *cosmovision* like that can feel the responsibility to save both human life and Mother Earth from destruction.

Web Sites

https://ourworldindata.org/extreme-poverty
www.thp.org/knowledge-center/know-your-world-facts-about-hunger-poverty/
www.who.int/news-room/detail/15-07-2019-world-hunger-is-still-not-going-down-after-three-years-and-obesity-is-still-growing-un-report
January 2020 Oxfam's report 'Time to Care': www.oxfam.org/en/press-releases/worlds-billionaires-have-more-wealth-46-billion-people
www.mayanleague.org/human-rights

The Calling of Global Responsibility
The Great Regeneration and the Great Responsibilization

Bian Li, The Hungry Lab, USA

I am honoured to be invited by Ananta Kumar Giri to contribute an afterword for this book, the *Calling of Global Responsibility: New Initiatives in Justice, Dialogues and Planetary Realizations.* This book comes timely as we, the larger society, find ourselves in the midst of a collective shift in consciousness as COVID-19 has brought to the forefront the harsh realities of a complacent world in the face of so many planetary challenges. The pandemic has also laid bare the systemic injustices, the structural inequalities, and the historical imbalances that have brought upon the consequences of our inaction. Yet most importantly, such times have accelerated our conversations, solutions, and movements forward—as individuals, communities and humanity itself—on our path towards a more enlightened society and in the process, a more inclusive, accessible, and just future.

The time is nigh for dialogue on such illuminating realizations, particularly those that inspire measurable action, and I congratulate Dr. Giri for his foresight and vision in advancing the global discourse with this book, as he explores the fundamental questions we must first ask, particularly on truth. For without truth, there can be no justice, and without justice, our realizations can only remain far-fetched ideas without purpose-driven implementation. As Alexis de Toqueville wrote in *Democracy in America*, at the dawn of a new experiment in social construct in the form of the fledgling USA, *"Men will not accept truth at the hands of their enemies, and truth is seldom offered to them by their friends."* Herein lies the great challenge of our increasingly polarized, factionalized and politicized world—how do we even begin to navigate the daunting journey of creating a more just society in the dawn of a post-COVID age when we cannot even agree on what the truth is? Where do we go and how do we get there if we cannot agree on where we are and how we got here?

At the heart of this conundrum lies the answer, and this book elevates the discussion to a global introspective by asking us to rethink and reconsider the meaning of global justice while transforming it to align with the larger planetary co-realizations through the lens of responsibilization.

I was recently asked to contribute an essay to the *Together* book,[38] an anthology of human experiences in the age of COVID-19. My essay, *The*

DOI: 10.4324/9780429347481-18

Great Regeneration of 2020, lays out what our collective experience will be looked back upon as a series of Greats:

> The **Great Awakening** to a new way of working, learning and being. Awakening to what's truly important—and essential—when this global pandemic has held up a mirror to our collective consciousness and reflected ourselves back to us—the good, the bad, the ugly. Awakening to the realization that so much of our modern world has been noisy, superficial, artificial and all-too fragile. Awakening to what's been repressed and forgotten within ourselves—our souls, our dreams, our identities beneath the artificial masks we show the world.
>
> The **Great Resetting** of Mother Nature—to heal, to replenish, to show us how much of a mess we've created and how important and interconnected she is to our own recovery. The clearing of polluted skies, the respawning of depleted fauna, the regrowth of damaged flora, with the urgent reminder that we are still too precipitously close to the point of no return.
>
> The **Great Reckoning** of the collective sins of our past. The long-overdue global reckoning of systemic racial inequalities, police brutality and historical injustices. The reckoning of our individual prejudices, microaggressions and subconscious biases. The reckoning of the pain caused by perpetrators, the protests harshly quelled and the silence of otherwise good men.

Above all, we are in the age—albeit just the nascent beginning—of The **Great Regeneration.** *Never before has there been such a sense of urgency to prepare society to adapt to a volatile tomorrow, with growing anxiety about being left behind. Building resilience is no longer enough. Resiliency is the ability to take the hits. Regeneration is a transcendence of those challenges, bending them to one's advantage and flourishing. Regeneration is a quiet revolution—breaking down the old normal of what led us into trouble and bringing forth a new normal that reshapes, redefines and reimagines a better future. A more inclusive, holistic and sustainable future for all. A regeneration that allows everyone, not just the privileged, the audacity to dream.*

With the arguments posited in Giri's book, we must add **The Great Responsibilization.** The collective shift brought on by our new planetary realizations compels us to review responsibility from a comparative global perspective, whose responsibility is it to lead the way? For what must we be responsible and accountable? By which actors to whom the burden is shouldered? And to whom should the solutions and advancements benefit? These questions serve as the guiding compass with which to navigate a greater clarity on morality, righting of historical wrongs and building a more inclusive future.

And in the process of answering such questions, we cannot ignore the significant role that corporations play, and in turn, the reciprocal role that consumers play, in driving forward new principles of sustainability, social, and spiritual responsibility. It is profound and prescient that Giri distinguishes between social and spiritual responsibility, though I disagree with the argument that corporations do not have souls. Legally, corporations are considered as people and operationally, are comprised of individuals who do, in fact, have souls. The aggregation of individual souls—be they healthy, happy, or suppressed—accordingly shape the collective corporate consciousness (or lack thereof) that drive the entity's actions—be they constructive or destructive, forward-thinking or short-sighted, progressive or regressive.

With growing consumer-savviness and awareness of sustainability and transparency in demanding greater accountability for corporate actions, ethical values, hiring practices, and environmental impact, an increasing number of corporations are now embedding, or at least promoting, ESG principles and criteria into company operations. Environmental, social, and governance (ESG) principles

> are a set of standards for a company's operations that socially conscious investors use to screen potential investments. Environmental criteria consider how accompany performs as a steward of nature. Social criteria examine how it manages relationships with employees, suppliers, customers, and the communities where it operates. Governance deals with a company's leadership, executive pay, audits, internal controls, and shareholder rights.[39]

While the actual effect and implementation of ESG principles vary widely and often are viewed with scepticism as corporate marketing or public reactions gimmicks, the fact that ESG is a rapidly growing topic of discussion taken more seriously than ever before in the corporate world is a direct reflection of the greater evolution from traditional **Shareholder Capitalism** to the new and much-needed **Stakeholder Capitalism**.[40] While shareholder capitalism has driven our modern economy in the insatiable quest for greater profits, markets, and customers at the expense of the environment, ethics, and employee well-being, stakeholder capitalism is aligning the needs of not only the corporation, its shareholders, and customers, but also its larger ecosystem of suppliers, communities, and environment who are impacted directly or as a by-product of the company's actions.

As explained by the World Economic Forum,

> The Fourth Industrial Revolution is forcing decision-makers to rethink how they create value and reinvent the ways in which their organizations function. . . . Future-oriented collaborative models require new corporate governance approaches that are much less based on traditional vertical control and siloed mechanisms while still maintaining accountability to shareholders. During COP25, the United Nations emphasized the critical importance of breakthrough innovation to achieve the Sustainable Development Goals by 2030. It is a matter of

corporate governance to put together the mechanisms that can reconcile how innovation can both enable long-term economic growth and fulfil each organization's purpose.

While currently gaining traction as a new philosophy whose time has come, especially in the age of COVID-19, the modern ethos behind widening the narrow spectrum of shareholder-based corporate governance and economic metrics to include sustainable benchmarks for human and environmental development has previously been espoused by forward-thinking leaders ahead of their time.

In a seminal speech at the University of Kansas in March 1968,[41] Robert F. Kennedy implored the audience to think beyond material measurements of wealth in order to elevate the importance of human dignity, social justice, and environmental stewardship:

> And this is one of the great tasks of leadership for us, as individuals and citizens this year. But even if we act to erase material poverty, there is another greater task, it is to confront the poverty of satisfaction—purpose and dignity—that afflicts us all. Too much and for too long, we seemed to have surrendered personal excellence and community values in the mere accumulation of material things. Our Gross National Product, now, is over $800 billion dollars a year, but that Gross National Product—if we judge the United States of America by that—that Gross National Product counts air pollution and cigarette advertising, and ambulances to clear our highways of carnage. It counts special locks for our doors and the jails for the people who break them. It counts the destruction of the redwood and the loss of our natural wonder in chaotic sprawl. It counts napalm and counts nuclear warheads and armored cars for the police to fight the riots in our cities. It counts Whitman's rifle and Speck's knife, and the television programs which glorify violence in order to sell toys to our children.

Yet the gross national product does not allow for the health of our children, the quality of their education or the joy of their play. It does not include the beauty of our poetry or the strength of our marriages, the intelligence of our public debate or the integrity of our public officials. It measures neither our wit nor our courage, neither our wisdom nor our learning, neither our compassion nor our devotion to our country, it measures everything in short, except that which makes life worthwhile. And it can tell us everything about America except why we are proud that we are Americans.

Such governance concepts are not new and the evolution of such a holistic, integrated, and sustainable form of corporate and economic governance and stewardship derives from a wholly spiritual concept, originating from the laws for a peaceful and harmonious society developed by many indigenous peoples around the world. One widely recognized concept is the Seventh

Generation Principle, based on an ancient Native American Haudenosaunee (Iroquois) philosophy that "the decisions we make today should result in a sustainable world seven generations into the future"[42] This concept was important and was codified in the Great Law of Haudenosaunee Confederacy, which formed the political, ceremonial, and social fabric of the Five Nation Confederacy (later Six). The Great Law of Haudenosaunee Confederacy is also credited as being a contributing influence on the American Constitution, due to Benjamin Franklin's great respect for the Haudenosaunee system of government.[43]

Many other indigenous peoples around the world have similar understandings of the inextricable linkages between man, nature, and the larger ecosystem, and even those peoples who have long-resisted modernization are now coming out from their ancestral homes to warn us about the consequences facing our facture if we continue ahead on our destructive path of divorcing capitalism from consciousness and corporate returns from corporate responsibility.

The ancient Arhuaco peoples of the Santa Maria Mountains of Colombia epitomize the harmony between the material and spirit realms in guiding environmental stewardship for future generations. Many traditional Arhuaco Mamos, or enlightened spiritual leaders, are now sharing their philosophies with modern society in an effort to change our ways. The Arhuaco believe that humanity exists within a universe that is wholly alive in and of itself, where all matter is embedded with life and consciousness. In this ancient "cosmovision" of the world, the Arhuaco maintain worship and stewardship of Mother Nature and "the thoughts of our ancestors are embedded in every rock and other element in which humans have contact. It's unfathomable to them that 'modern man' does not believe the Earth consciously experiences the harm we inflict on it."[44]

In their vernacular, we modern societies are referred to as "Younger Brother" in need of guidance while they are the "Elder Brother" whose wisdom must be deployed in these urgent times to stem the tide of climate change. In the words of an Arhuaco mamo, "the Younger Brother is damaging the world. He is on the path to destruction. He must understand and change his ways, or the world will die."[45]

Whether ancient or modern, corporate or cosmic, economic or environmental, it cannot be denied that for all our conflicts, challenges and polarities, we are, in the end, individual parts of the whole. And we cannot be responsible only for ourselves without affecting the whole. In ancient times, it was written in the Upanishads books of the Vedas "*Yatha PindeTatha Brahmande.*" As is the atom so is the universe. Rumi agreed: "You are not a drop in the ocean. You are the ocean in every drop." And in modern times, Einstein opined:

> I like to experience the Universe as one harmonious whole. Every cell has life. Matter, too, has life; it is energy solidified. . . . The soul given to each of us is moved by the same living spirit that moves the Universe.

And it is from this truism of nature that we must advance forward in the pursuit of a more just, enlightened future.

It is this elevated cognizance that our responsibility to ourselves and our responsibility to our larger humanity are integrally connected that will accelerate the shift from, as I argue in my work, the next industrial revolution (the future of doing) to the next Cognitive Revolution (the Future of Learning) into the ultimate transformation, and the inevitable destination of the human experience—the next Existential Revolution (the Future of Being) as we evolve our understanding of who we are and take our place in an uncertain and volatile tomorrow.

I have written and spoken much on the interplay and duality of science and spirituality (not to be confused with religion), which can be argued are just two sides of the same coin. By adding Corporate Spiritual Responsibility to CSR, Giri creates a safe space to challenge existing paradigms and inspire us to think deeper about the true purpose of why, how and for whom our influential organizations can impact society at large. While the concept of spirituality within a corporate setting can often be seen as frivolous, dogmatic or "new age," the greatest scientists of our time understood the ineffable yet pervasive entanglement between our material world with the greater forces that drive our wonderment, exploration and intentionality with which we shape our destinies.

As explained by Einstein:

> If we look at this tree outside whose roots search beneath the pavement for water, or a flower which sends its sweet smell to the pollinating bees, or even our own selves and the inner forces that drive us to act, we can see that we all dance to a mysterious tune, and the piper who plays this melody from an inscrutable distance—whatever name we give him— Creative Force, or God—escapes all book knowledge. . . . [A cosmic religion] has no dogma other than teaching man that the Universe is rational and that his highest destiny is to ponder it and co-create with its laws.[46]

It is with this humility—the acceptance that we are not the masters of the universe but rather its subjects, and that we cannot escape the machinations of our own demise unless we embrace a rethinking and transformation of the meaning of responsibility and justice—that we must venture forth. And that is at the crux of Giri's book as he elevates the dialogue through this broader lens.

It is in this spirit that I leave you with a final introspection on the journey of us—who we are, where we are, how we got here, where we want to go, and ultimately, how we can get there. In 1994, the late renowned astrophysicist and philosopher Carl Sagan gave a speech called "The Pale Blue Dot," narrating to us his didactic thoughts on our dear planet Earth, captured as an infinitesimal dot in a snapshot from the Voyager I spacecraft. In all my years of work as a futurist, educationist, and social innovation expert, I have yet to encounter a greater definition more poignantly put than the words in his iconic speech[47] reminding us of our ultimate, and truest, call to global responsibility through the highest form of planetary realization:

> Look again at that dot. That's here. That's home. That's us. On it everyone you love, everyone you know, everyone you ever heard of, every human being who ever was, lived out their lives. The aggregate of our joy and suffering, thousands of confident religions, ideologies, and economic doctrines, every hunter and forager, every hero and coward, every creator and destroyer of civilization, every king and peasant, every young couple in love, every mother and father, hopeful child, inventor and explorer, every teacher of morals, every corrupt politician, every "superstar," every "supreme leader," every saint and sinner in the history of our species lived there—on a mote of dust suspended in a sunbeam.
>
> The Earth is a very small stage in a vast cosmic arena. Think of the rivers of blood spilled by all those generals and emperors so that, in glory and triumph, they could become the momentary masters of a fraction of a dot. Think of the endless cruelties visited by the inhabitants of one corner of this pixel on the scarcely distinguishable inhabitants of some other corner, how frequent their misunderstandings, how eager they are to kill one another, how fervent their hatreds.
>
> Our posturing, our imagined self-importance, the delusion that we have some privileged position in the Universe, are challenged by this point of pale light. Our planet is a lonely speck in the great enveloping cosmic dark. In our obscurity, in all this vastness, there is no hint that help will come from elsewhere to save us from ourselves.
>
> The Earth is the only world known so far to harbor life. There is nowhere else, at least in the near future, to which our species could migrate. Visit, yes. Settle, not yet. Like it or not, for the moment the Earth is where we make our stand.
>
> It has been said that astronomy is a humbling and character-building experience. There is perhaps no better demonstration of the folly of human conceits than this distant image of our tiny world. To me, it underscores our responsibility to deal more kindly with one another, and to preserve and cherish the pale blue dot, the only home we've ever known.

Global Responsibility and the Calling of Planetary Realizations

Walking and Meditating Together With Epilogues and Reflections

Ananta Kumar Giri

> That under conditions of terror most people will comply but some people will not. . . . Humanly speaking, no more is required, and no more can reasonably be asked, for this planet to remain a place fit for human habitation.
> Hannah Arendt (1976), *Eichmann in Jerusalem*

> Nationalist isolationism is probably even more dangerous in the context of climate change than of nuclear war.
> Yuval Noah Harari (2018), *21 Lessons for the 21st Century*

> The fundamental problem, I believe, is that at every level we are giving too much attention to the external, maternal aspects of life while neglecting moral ethics and inner values.
>
> By inner values I mean the qualities that we all appreciate in others, and toward which we all have a natural instinct, bequeathed by our biological nature as animals that survive and thrive only in an environment of concern, affection, and warm-heartedness—or in a single word, compassion. The essence of compassion is a desire to alleviate the suffering of others and to promote their well-being. This is the spiritual principle from which all other positive inner values emerge. We all appreciate in others the inner qualities of kindness, patience, tolerance, forgiveness and generosity, and in the same way we are all averse to displays of greed, malice, hatred and bigotry. So actively promoting the positive inner qualities of the human heart that arise from our disposition toward compassion, and learning to combat our more destructive propensities, will be appreciated by all. And the first beneficiaries of such strengthening of inner values will, no doubt, be ourselves. Our inner values are something we ignore at our own peril, and many of the greatest problems we face in today's world are the result of such neglect.

Dalai Lama (2011), "Compassion and the Question of Justice," pp. x–xi
> "Opening up to the world" is an expression that has been co-opted by the economic and financial sector and is now used exclusively of openness to foreign interests or to the freedom of economic powers to invest without obstacles or complications in all countries. Local conflicts and disregard for the common good are exploited by the global economy in order to impose a single cultural model. This culture unifies the world, but divides persons and nations, for "as society becomes ever more globalized, it makes

DOI: 10.4324/9780429347481-19

us neighbours, but does not make us brothers". We are more alone than ever in an increasingly massified world that promotes individual interests and weakens the communitarian dimension of life.

[B]y acknowledging the dignity of each human person, we can contribute to the rebirth of a universal aspiration to fraternity. Fraternity between all men and women.

Pope Francis (2020), *Fratelli Tulti: On Fraternity and Social Friendship*

I am grateful to Professor Fred Dallmary, P.V. Rajagopal, Dr. Felix Padel, Dr. A. Osman Farah, Professor Jeffrey Haynes, Julie M. Geredien, Dr. Sapir Handelman, Dr. Rico Sneller, Professor Sabelo J. Ndlovu-Gatsheni, Dr. Mahmoud Masaeli, and Bian Li for joining us in this collaborative exploration of the calling of global responsibility cultivated in this book. They bring many rich new insights and help us in our visions and practices of responsibility. There are many criss-crossing resonant themes and co-insights in their reflections which call for further walking and meditating together with these thoughts and practices.

In his epilogue, Dallmayr tells us:

The new world which is emerging—and Giri's book allows us to peak into the process—is indeed an un-habitual world, one not firmly tied down to traditional structures or conventions. The ongoing process is still fluid or in flux and does not allow for neat definitions.

For Dallmayr, the book helps us in envisioning and practising global humanization. For Dallmayr, "the chapters of the book do not offer a complete picture of the emerging 'New World' (which would be quite impossible today), but they all point intriguingly in the new direction." Dallmayr invites us to realize that

it is by no means clear or predictable whether the agents of the "new world" will be less susceptible to corruption, domination and violence than the agents of the "old." This question cannot be resolved by scholars in their books. All depends on proper education, exhortation, and guidance by good example—which is the global responsibility of us all.

Dallmayr tells us about the calling of being examples of responsibility and PV Rajagopal is an example of this. We have discussed about his work and vision and that of Ekta Parishad in Chapter 2. Among many insights that he shares in his epilogue is the following: "I hope the day will come when we can introduce a new theory of 'making everyone fit for survival' rather than 'survival of the fittest.'" Rajagopal also emphasizes the significance of putting a responsible global governance system in place. For Rajagopal,

The adverse effects of the COVID-19 pandemic have taught us what can happen if we don't have responsible global governance system in place. It is time to decentralize much of the governance responsibility to the communities at the local level and also to give up part of the responsibility to enable an effective global governance system based on consultation and collaboration.

For Rajagopal,

Issues like poverty, inequality, the climate crisis, warmongering or even this pandemic, cannot be tackled by any one country. It demands a responsible global governance system to deal with the serious issues of wide-spread poverty; distress migration; large-scale human rights violations, as these cannot be left to whimsical and dangerous leaders at the national levels. This is the time to look forward and develop forms of governance that give people hope for the future rather than being paralyzed by lethargy.

Rajagopal urges us to realize how important it is "to gauge the capacity of non-state actors in influencing and directing responsible global governance."

Rajagopal tells all of us to have a responsibility to engage creatively and critically at both local and global levels. What Rajagopal tells us about Ekta Parishad deserves our careful consideration:

Though we began as an organisation addressing local issues, as time went on, we grew to realize that we needed to play both at the local and global levels, if any meaningful change were to be brought about during our lifetime.

Resonating with the spirit of the book, Rajagopal tells us:

Analogous to the arguments of Giri, this effort has to build a global justice movement rather than anti-globalisation movement. The Jai Jagat campaign has set a positive agenda where there is a greater chance for individuals and organisations to come together.

This resonates with the local plus and globalization plus perspectives of Bruno Latour discussed in our book.

Rajagopal's epilogue is accompanied by that of Felix Padel who emphasizes the significance of environmental justice and global well-being. Padel here tells us how we can learn from indigenous and primal communities of the world such as the Adivasis of India about environmental care and community and Earth well-being. Padel engages with the presentation of the work of *Roots* in chapter Three and invites us to cultivate dual perspectives in situations of conflicts which can help us in cultivation of responsibility. As Padel writes:

This book highlights the vital dialogue between Israelis and Palestinians promoted by "Roots" (in Chapter 2). We need dual narratives of how Israelis were killed and pushed out of their homes throughout Europe, North Africa and the Middle East, alongside stories of the Nakba about how Palestinians have been displaced by Israeli settlers, starting in 1948, exactly when the UN Charter of Human Rights was released, and continuing in 2020. We need similar dual narratives on Kashmir, by Kashmiri Pandits alongside mention of "disappearances" and stories of Muslim inhabitants who have lost loved ones to violence by security forces and others. We need similar dual narratives of the terrible ethnic cleansing that swept Greece and Turkey, resulting in the forced exchange of populations in the 1920s; in former Yugoslavia, and many other regions, until dialogue transforms "enemies" into friends everywhere.

Global responsibility calls for not only dual narratives but also mutual and multiple narratives.

Padel's afterword is followed by that A. Osman Farah who in his epilogue invites us to recover justice from irresponsibility to responsibility. Farah builds upon the insight from the book that "justice refers not merely to the structural criteria of producing universal social and political goods." "Rather justice often occurs in a non-static dynamic process emerging, not from pre-given strategic framing, but situations in which people imagine, perceive, discuss, and intersubjectively project uni-dimensional as well as multi-dimensional reasoning." Farah states that the book invites us to look at justice as a transcendental "unfinished project" and realize "justice as responsibility." Farah presents works in transnational justice, cross-cultural and inter-generational communication in Aarhus, Denmark especially between local Danes and Somalis where "justice is not just interpersonal or inter-group but also concerns intergenerational relations that remain the most vital/critical human relations. In the end therefore, what matters could be the transfer/transmission of inclusive knowledge and wisdom to future generations." Farah's reflections are accompanied by Jeffrey Haynes' who tells us about the significance of dialogues among religions as well climate emergency for the realization of global responsibility. Haynes writes:

> What is my perspective on dialogue and global responsibility and what are the challenges facing us all in these respects? The planet is beset with very significant challenges. It is invidious to try to place them in a hierarchy of concern. From my perspective, however, the immense challenge of the climate emergency stands out as both the most imminent challenge and the one in which we all have key responsibility. We have personal responsibility to treat the panel better, and we also have a collective responsibility. What is my perspective on dialogue and global

responsibility and what are the challenges facing us all in these respects? The planet is beset with very significant challenges. It is invidious to try to place them in a hierarchy of concern. From my perspective, however, the immense challenge of the climate emergency stands out as both the most imminent challenge and the one in which we all have key responsibility. We have personal responsibility to treat the planel better, and we also have a collective responsibility.

In her epilogue, Julie M. Geredien discusses further about International Charter of Responsibility and highlights the work of Edith Sizoo and the Charles Leopold Mayer Foundation for Human Progress. Thus Geredien deepens our understanding of global responsibility. Geredien's is followed by Sapir Handelman's who tells us about the responsibility of finding ways out of intractable conflicts such as between Israel and Palestine. Handelman tell us about citizen initiatives such as The Congress of the People which is a Public *Negotiating* Congress. As a philosopher and peace maker, Handelman has given leadership in this effort in the midst of the Corona Crises which tells us how much each one of us can do for taking one and many steps together for the realization of global responsibility. Handelman's epilogue is followed by Sabelo J. Ndlovu-Gatsheni's who invite us to understand the significance of decolonial turn, especially onto-decolonial turn, for cultivation of global responsibility (also see Chimakonam 2017). Rico Sneller then joins us in introducing what he calls "four indispensable moral notions . . . 'retrospection,' 'recollection,' 'memory,' and 'forgiveness'" to our vision and practice of global responsibility. Sneller tells us:

> For example, the problem of "global inequality"—which has been discussed extensively in Giri's book—would be handled inadequately if it were only seen as a subject matter that can be clearly defined in terms of objectifiable *external* phenomena—as if global inequality were an "object" for a "subject". Instead, "global inequality" (to stick to this example) should first and foremost be framed as a form of growing self-*alienation* in those defining and addressing it before being addressed by it.

Coming from Brazil and with close collaboration with social and cultural movements in South America, Beatriz Bassio in her reflections emphasizes the significance of alternative visions of the world such as PachaMama and Gaia for realization of global responsibility. For Bassio,

> The 18th and 19th century model of the world as a machine has led humans to the vocation to control nature; on the contrary, thinking of the Earth as Gaia, as Mother Earth, as a web of relationships between our natural habitat and life in all its manifestations leads to the responsibility to take care for it. It is a change of paradigm.

Such a view is presented to us not only by James Lovelock's understanding of our Earth as Gaia but also by the cosmovision of the Mayans who invite us to realize our Earth as PachaMama. Mayan philosophy tells us that "*Cosmovision* and spirituality are a simultaneous experience, acting at the same time, myth and history, death and resurrection." For Bassio, "Only a human civilization with a *cosmovision* like that can feel the responsibility to save both human life and Mother Earth from destruction." In her concluding epilogue to this conversation, Bian Li, a creative poetess and futurist activist, invites us to bring great awakening, great resetting, great reckoning, and great regeneration to our current moment of great acceleration. Li invites us to bring science and spirituality together our strivings (see Capra 1984).

Thus, our interlocutors enrich our visions and practices of global responsibility from their diverse backgrounds of knowledge and experience. We join them in crafting our own visions and practices of responsibility—self and social; local, global, and the planetary. We deepen our own roots, cross-fertilize our roots, and fly with our new wings of actions, imaginations, and meditations. As Sri Aurobindo calls (1993) us in his epic *Savitri*:

> [A]nd they shall turn to meet a nameless tread,
> Adventurers into a mightier Day.
> Ascending of the limiting breadths of mind,
> They shall discover the world's huge design
> And step into the Truth, the Right, the Vast
> You shall reveal to them the hidden eternities,
> The breath of infinitudes not yet revealed.

Notes

1 In his theory of metamorphosis, Urlich Beck talks about "organized irresponsibility" in our contemporary world (see Mythen 2018).
2 This relates to Strydom's concepts of functional and communicative globalization.
3 For a detailed discussion on the characteristics and symptoms of intractable conflict, see, for example, Kriesberg (1993); Bar-Tal (1998); Coleman (2000); Kelman (2007).
4 One of the classical symptoms of intractable conflict is—any peacemaking progress leads to a significant increase in the level of violence. The people need to be involved in the peacemaking process in order to cope with the destructive implications of this symptom. For a further discussion, see Handelman (2016).
5 For a comprehensive account of the South African case, see Sparks (1994). For a comprehensive account of the Northern-Ireland case, see Mitchell (1999).
6 The initiative grows from the Minds of Peace Experiment—a brief Israeli-Palestinian Public Negotiating Assemblies—which were conducted in different formats and locations. For a further discussion, see, for example, Handelman (2010) and Handelman and Pearson (2014: 25–19). For more details on the project, visit mindsofpeace.org
7 The meeting was covered by the Australian Public TV, SBS: www.youtube.com/watch?v=U4I7Jt9BCpg&feature=youtu.be&fbclid=IwAR0u3Nm2_fsAJbnZ3 JQOlvWbBh6doCZOTJahrjKlFjH0_EST-an_ORp1sZQ.

8 One of the major events following the Israeli unilateral disengagement from Gaza strip was a coup d'état of Hamas that led to a political split in the Palestinian society: the radical Islamist movement, Hamas, controls Gaza; the secular nationalist movement, Fatah, administrates The West Bank.
See Huntington (2006).

9 Cesaire, A. 1955 [2000]. *Discourse on Colonialism: Translated by Joan Pinkham.* New York: Monthly Review Press, p. 31.

10 Ndlovu-Gatsheni, S. J. 2020. *Decolonization, Development and Knowledge in Africa: Turning Over a New Leaf.* New York and London: Routledge; Ndlovu-Gatsheni, S. J. 2020. 'The Cognitive Empire, Politics of Knowledge and African Intellectual Productions: Reflections on Struggles for Epistemic Freedom and Resurgence of Decolonization in the Twenty-First Century.' *Third World Quarterly*, pp. 1–20.https://doi.org/10.1080/01436597.2020.17775487.

11 Thiongo, Ngugi wa. 1993. *Moving the Centre: The Struggle for Cultural Freedom.* Oxford: James Currey, pp. xvi–xvii.

12 Reiter, B. (ed.). 2018. *Constructing the Pluriverse: The Geopolitics of Knowledge.* Durham and London: Duke University Press; Mignolo, W. D. and Walsh, C. E. 2018. *Decoloniality: Concepts, Analytics, Praxis.* Durham and London: Duke University Press; and Escobar, A. 2018. *Designs for the Pluriverse: Radical Interdependence, Autonomy, and the Making of Worlds.* Durham and London: Duke University Press.

13 Thiong'o, Ngugi wa. 2012. *Globalectics: Theory and the Politics of Knowing.* New York: Columbia University Press, p. 8.

14 Fanon, F. 1968. *The Wretched of the Earth.* New York: Grove Press, p. 252.

15 Blaut, J. 1993. *The Colonizer's Model of the World: Geographical Diffusionism and Eurocentric History.* New York: The Gilford Press. See also Ndlovu-Gatsheni, S. J. 2020. *Decolonization, Development and Knowledge in Africa: Turning Over a New Leaf.* London and New York: Routledge.

16 Makoni, E. N. and Ndlovu-Gatsheni, S. J. 2020. 'Redesigning the World: A Decolonial View from the Epistemologies of the Global South.' *KERB: Journal of Landscape Architecture*, 28, forthcoming.

17 Mignolo, Local Histories/Global Designs.

18 Ndlovu-Gatsheni, S J. 2018. *Epistemic Freedom in Africa: Deprovincialization and Decolonization.* New York and London: Routledge.

19 Blaut, J.M. 1993. *The Colonizer's Model of the World.* New York: The Guildford Press, p. 1.

20 Quijano, A. 2000. 'The Coloniality of Power and Social Classification.' *Journal of World Systems*, 6(2) (Summer-Fall), pp. 342–386.

21 Ndlovu-Gatsheni, S.J. 2013. *Empire, Global Coloniality and African Subjectivity.* Oxford and New York: Berghahn Books.

22 Santos, B. de S. 2007. 'Beyond Abyssal Thinking: From Global Lines to Ecologies of Knowledges', *Review*, 30(1) at 1.

23 Grosfoguel, R. 2007. 'The Epistemic Decolonial Turn: Beyond Political-Economy Paradigms.' *Cultural Studies*, 21(2–3) (March/May), pp. 211–223.

24 Santos, B. de. S. 2007. 'Beyond Abyssal Thinking: From Global Lines to Ecologies of Knowledges.' *Review*, 30(1), pp. 45–89.

25 Maldonado-Torres, N. 2007. 'On the Coloniality of Being: Contributions to the Development of a Concept.' *Cultural Studies*, 21(2–3)(March/May) at 255.

26 Quijano, A. 2000. 'The Coloniality of Power and Social Classification.' *Journal of World Systems*, 6(2) (Summer-Fall), pp. 342–386.

27 These 'Ten Ds' of decolonization were formulated by Ndlovu-Gatsheni, S. J. 2020. *Decolonization, Development and Knowledge in Africa: Turning Over a New Leaf.* New York and London: Routledge; Ndlovu-Gatsheni, S. J. 2020. 'The Cognitive Empire, Politics of Knowledge and African Intellectual Productions:

Reflections on Struggles for Epistemic Freedom and Resurgence of Decoloniza-tion in the Twenty-First Century.' *Third World Quarterly*, pp. 1–20.https://doi.org/10.1080/01436597.2020.17775487.

28 Santos, B. de. S. 2007. 'Beyond Abyssal Thinking: From Global Lines to Ecologies of Knowledges.' *Review*, 30(1), pp. 45–89.

29 Fisher, W. and Ponniah, T. (eds.). 2015. *Another World Is Possible: World Social Forum Proposals for an Alternative Globalization*. London: Zed Books.

30 Cf Masaeli, M. and Sneller, R. (eds.). 2020. *The Return of Ethics and Spirituality in Global Development*. Antwerpen: Gompel&Svacina.

31 Interestingly, Biblical Hebrew does not really have an equivalent for the English verb "to ask." The first ethical "question" in the Bible is "said" rather than "asked": "And the LORD God called unto the man, and said [וַיֹּאמֶר] unto him: 'Where art thou?'" (*Genesis* 3, 9). Even reproach is "said": "And He said [וַיֹּאמֶר]: 'What hast thou done?'" (*Genesis* 4, 10).

32 Friedmann, Hermann. 1930/1925. *Die Welt der Formen. System eines morphologischen Idealismus*. München: C.H. Beck'sche Verlagsbuchhandlung, p. 499.

33 The distinction between thin and thick language originates in Gilbert Ryle and Bernard Williams.

34 https://ourworldindata.org/extreme-poverty. This underscores the immense challenge of achieving the Sustainable Development Goal of Zero Hunger by 2030, says a 2019 edition of the annual The State of Food Security and Nutrition in the World report. Ninety-ninepercent of the world's undernourished people live in developing countries. Where is hunger the worst? **Asia:** 513.9 million; **Sub-Saharan Africa:** 239.1 million; **Latin America:** 34.7 million www.thp.org/knowledge-center/know-your-world-facts-about-hunger-poverty/
      www.who.int/news-room/detail/15-07-2019-world-hunger-is-still-not-going-down-after-three-years-and-obesity-is-still-growing-un-report

35 January 2020 Oxfam's report. 'Time to Care'. www.oxfam.org/en/press-releases/worlds-billionaires-have-more-wealth-46-billion-people.

36 García Nogales, Mila. 2020. Ecofeminismo Más allá de Greta Thunberg: voces de mujeres indígenas en la lucha medioambiental, Público, Madrid, 26 June.

37 *Cosmovision* is a concept specifically related to a view of the world as understood by Latin American Indigenous cultures. It is similar to "worldview," "cosmology," or "Weltanschaung," in German.

38 Together: An anthology from the COVID-19 pandemic, Aditya, Otermans, 2020.

39 www.investopedia.com/terms/e/environmental-social-and-governance-esg-criteria.asp

40 www.weforum.org/agenda/2020/01/shift-to-stakeholder-capitalism-is-up-to-us/

41 www.jfklibrary.org/learn/about-jfk/the-kennedy-family/robert-f-kennedy/robert-f-kennedy-speeches/remarks-at-the-university-of-kansas-march-18-1968

42 www.ictinc.ca/blog/seventh-generation-principle

43 www.ictinc.ca/blog/seventh-generation-principle

44 www.bbc.com/travel/story/20190329-the-ancient-guardians-of-the-earth

45 www.bbc.com/travel/story/20190329-the-ancient-guardians-of-the-earth

46 Einstein and the Poet: In Search of the Cosmic Man (1983). From a series of meetings William Hermanns had with Einstein in 1930, 1943, 1948, and 1954.

47 Carl Sagan, Pale Blue Dot, 1994.

References Cited

Abdurrahman Wahid & Daisaku Ikeda. 2015. *The Wisdom of Tolerance: A Philosophy of Generosity and Peace*. London: I.B. Tauris.

Agamben, Giorgio. 1995. *Homo Sacer: Sovereign Power and Bare Life*. Stanford: Stanford U. Press.

Akinade, Akintunde E. 2014. *Christian Responses to Islam in Nigeria: A Contextual Study of Ambivalent Encounters*. New York: Palgrave Macmillan.

Ankersmit, Frank. 1996. *Aesthetic Politics: Political Philosophy Beyond Fact and Value*. Stanford: Stanford U. Press.

Antentas, Josep Maria. 2017. "Spain: From the Indignados Rebellion to Regime Crisis (2011–2016)." *Labor History* 58 (1): 106–131. DOI: 10.1080/0023656X.2016.12 39875

Alexander, Jeffrey, Ron Eyerman, Bernhard Giesen, and Neil J. Smelser. 2004. *Cultural Trauma and Collective Identity*. Berkeley: University of California Press.

Apel, Karl-Otto. 2000. "Globalization and the Need for Universal Ethics." *European Journal of Social Theory* 3 (2): 137–155.

Appadurai, Arjun. 2013. *Future as a Cultural Fact*. London: Verso.

———. 2020. "Corona Virus Won't Kill Globalization But It Will Look Different after the Pandemic." *TIME*. May 19.

Arendt, Hannah. 1976. *Eichmann in Jerusalem: An Essay on the Banality of Evil*. New York: Viking.

———. 1988 [1958]. *The Human Condition*. Chicago: University of Chicago Press.

Baker, Katherine. 2018. "An Education for Responsibility: Edith Stein and the Formation of the Whole Person." In *Justice and Responsibility: Cultural and Philosophical Foundations*, (eds.) Joao J. Vila-Cha and John P. Hogan, pp. 159–170. Washington, DC: The Council for Research in Values and Philosophy.

Bar-Tal, Daniel. 1998. "Societal Beliefs in Times of Intractable Conflict: The Israeli Case." *International Journal of Conflict Management* 9 (1): 22–50.

Bartolf, Christian. 2018. "Music and Cosmopolitanism. In *Beyond Cosmopolitanism*, (ed.) Ananta Kumar Giri. New York: Palgrave Macmillan.

Bateson, Gregory. 1972. *Steps to an Ecology of Mind: Essays in Anthropology, Psychiatry, Evolution, and Epistemology*. Chicago: University of Chicago Press.

Bauman, Zygmunt. 2013. *Does the Richness of the Few Benefit Us All?* Cambridge: Polity Press.

Baxi, Upendra. 2020. "Exodus Constitutionalism." *India Forum*. July 3.

Beck, Ulrich. 1992. *Risk Society: Towards a New Modernity*. London: Sage.

Bellah, Robert N. et al. 1991. *The Good Society*. New York: Alfred A Knof.

Berger, Julia. 2003. "Religious Nongovernmental Organisations: An Exploratory Analysis." *Voluntas* 14 (1): 15–39.

Berger, Peter (ed.). 1999. *The Desecularization of the World: Resurgent Religion in World Politics*. Grand Rapids: William B. Eerdmans.

Bernsten, Monica Sydgard. 2002. *ATTAC Norway: An Anthropological Study of the Norwegian Branch of Transnational Social Movement*. Master's Theses. Department of Social Anthropology, University of Oslo.

Bharati, Acharya Sachidananda. 2017. *The Air Plot: Socio-Spiritual Foundations of Integral Revolution*. Kottayam: Winco Books.

Bhatt, Ela. 2015. *Anubandh: Building Hundred Mile Communities*. Ahmedabad: Navjivan Trust.

Bourdieu, Pierre. 1991. *The Political Ontology of Martin Heidegger*. Cambridge: Polity Press.

Boyd, Thomas. 2017. *Listening Projects for the Self and Social Action*. Theses. SIT Graduate Institute. Brattleboro, Vermont.

Buber, Martin. 1958. *I and Thou* . New York: Scribner.

Butler, Judith. 2020. *The Force of Non-Violence: The Ethical in the Political*. London: Verso.

———. 2021 [1997]. *Excitable Speech*. London: Routledge.

Capra, Fritjof. 1984. *The Turning Point: Science, Society, and the Rising Culture*. New York: Bantam.

Carrette, Jeremy, and Hugh Miall. 2013. *Religious NGOs and the United Nations*. New York, Geneva and Canterbury: University of Kent.

Carrette, Jeremy, and Hugh Miall (eds.). 2018. *Religion, NGOs and the United Nations*. London: Bloomsbury Academic.

Carr-Harris, Gillian Sankey. 2021. *A Study on Non-Violence: Constructing Narratives of Leadership*. PhD Thesis. Ontario Institute for Study in Education, University of Toronto.

Chakraborty, Dipesh. 2019. *The Crises of Civilization: Exploring Global and Planetary Histories*. Delhi: Oxford U. Press.

———. 2020. "How the Pandemic Expands Our Vision of History." Inaugural Lecture at the International Webinar Series, "Writing Post-Pandemic Life World: Society, Cultural Materialities and Practices", Department of History, Ravenshaw University. July 8.

———. 2021. *The Climate of History in a Planetary Age*. Chicago: University of Chicago Press.

Chapple, Christopher. 2020. "The World's Religions in a Time of Pandemic." *ISJS-TRANSACTIONS: A Quarterly Referred Online Journal on Jainism* 4 (2): 1–6, April–June.

Chatterjee, Margaret. 2005. *Gandhi and the Challenge of Religious Diversity: Religious Pluralism Revisited*. New Delhi and Chicago: Promilla & Co.

Cheng, C. Y. 1989. "On Harmony as Transformation: Paradigms for the I Ching." *Journal of Chinese Philosophy* 16 (2): 125–158.

Chimakonam, Jonathan O. 2017. "African Philosophy and Global Epistemic Justice." *Journal of Global Ethics* 13 (2): 120–137.

Christian, David. 2004. *Maps of Time: An Introduction to Big History*. Berkeley: University of California Press.

Christians, Clifford G. 2008. "The Challenge of Responsibility in the United States: Individualism in the Face of Collective Obligation." In Sizoo: 65–80.

Ciudadanos. 2015. Programa Electoral. 20 December. http://servicios.lasprovincias. es/documentos/programa-electoral-ciudadanos-20D-2015.pdf

Clammer, John. 2016. *Cultures of Transition and Sustainability: Culture after Capitalism*. New York: Palgrave Macmillan.

Clammer, John, and Ananta Kumar Giri (eds.). 2017. *The Aesthetics of Development: Art, Culture and Social Transformations*. New York: Palgrave Macmillan.

Clooney, Frank. 2018. *Learning Interreligiously: In the Text, in the World*. Minneapolis: Fortress Press.

Coleman, Peter. 2000. "Intractable Conflict." In *The Handbook of Conflict Resolution: Theory and Practice*, (eds.) Morton Deutsch and Peter Coleman. San Francisco, CA: Jossey-Bass Publishers.

Collins, Peter M. 2018. "The Search of the 'Responsible Life' in Martin Buber and Leo Buscaglia." In *Justice and Responsibility: Cultural and Philosophical Foundations*, (eds.) Joao J. Vila-Cha and John P. Hogan, pp. 377–400. Washington, DC: The Council for Research in Values and Philosophy.

Connolly, William E. 2013. *The Fragility of Things: Self-Organizing Processes, Neoliberal Fantasies and Democratic Activism*. Durham: Duke U. Press.

Critchley, Simon. 2020. "To Philosophize Is to Learn How to Die." *New York Times*. April 11.

Dalai Lama. 2011. "Compassion and the Question of Justice." In idem, *Beyond Religion: Ethics for a Whole World*. Boston: Houghton Mifflin Harcourt.

Dallmayr, Fred. 2002. *Dialogue among Civilizations*. New York: Palgrave Macmillan.

———. 2007. "Love and Justice: A Memorial Tribute to Paul Ricouer." In idem, *In Search of the Good Life: A Pedagogy for Troubled Times*. Lexington: University of Kentucky Press.

———. 2013. *Being in the World: Dialogue and Cosmopolis*. Lexington: University of Kentucky Press.

Dandekar, Ajay, Felix Padel, and Jeemol Unni. 2014. *Ecology Economy: Quest for a Socially Informed Connection*. Hyderabad: Orient Blackswan.

Das, Chitta Ranjan. 2008. "Jagatikarana: Sanskrutika Parichiti" [Globalization and Cultural Identity]. In idem, *Manaku Stree besa Kari* [Adorning One's Mind as A Woman]. Bhubaneswar: Pathika Prakashani.

———. 2020. *The Essays of Chitta Ranjan Das on Literature, Culture and Society: ON the Side of Life in Spite of*. New Castle upon Tyne: Cambridge Scholars Press.

Das, Jishnu. 2020. "India's Response to Corona Virus Can't be Based on Existing Epidemiological Models." *The Print*. 6 April.

Das, Veena. 2003. "Introduction." *Oxford India Companion to Sociology and Anthropology*, (eds.) Veena Das. Delhi: Oxford University Press.

———. 2015. *Affliction: Health, Disease, Poverty*. New York: Fordham U. Press.

———. 2020a. "Facing Covid-19: My Land of Neither Hope Nor Despair." *American Ethnologist*. In *Covid-19 and Student-Focused Concerns: Threats and Possibilities*, (eds.) Veena Das and Naveeda Khan. American Anthropologist website. May 1.

———. 2020b "What Does Ordinary Ethics Look Like?" In *Four Lectures on Ethics: Anthropological Perspectives*, (eds.) Michel Lambek et al. Chicago: HAU Books.

———. 2020c "Corona Policy Must Factor in Scientific Uncertainty." *Deccan Herald*. May 24.

Demeterio 111, Feorillo. 2018. "Young's Theory of Structural Justice and Collective Responsibility." In *Justice and Responsibility: Cultural and Philosophical Foundations*, (eds.) Joao J. Vila-Cha and John P. Hogan, pp. 377–400. Washington, DC: The Council for Research in Values and Philosophy.

de Neeve, Geet. 2009. "Power, Inequality and Corporate Social Responsibility: The Politics of Ethical Compliance in the South Asian Garment Industry." *Economic and Political Weekly* 45 (22): 63–71.

Derrida, Jacques. 1994. *Specters of Marx*. London: Routledge.

de Sousa Santos, Boaventura. 2005. "Beyond Neo-Liberal Governance: The World Social Forum as Subaltern Cosmopolitan Politics and Legality." In *Law and Globalization from Below: Towards a Cosmopolitan Legality*, (eds.) Boaventuara de Sousa Santos and Cesar A. Rodriguez-Garanto, pp. 29–63. Cambridge: Cambridge U. Press.

———. 2008. "The World Social Forum and the Global Left." *Politics and Society* (2).

———. 2014. *Epistemologies of the South: Justice Against Epistemicide*. Boulder, CO: Paradigm Publishers.

Devy, Ganesh. 2010. *People's Linguistic Survey of India*. Hyderabad: Orient Blackswan.

Dixon-Decleve, Sandrine, Owen Gaffrey, Jayati Ghosh, Jorgen Randers, Johan Rockstrom, and Per Espen. 2022. *Earth for All: A Survival Guide for Humanity*. A Report to the Club of Rome (2022), Fifty Years After The Limits to Growth (1972). Gabriola Island, BC, Canada: New Society Publishers.

Doran, Marie Christine. 2014. "Religion and Land Takeovers in Mexico: Collective Miracle Discourses and the Building of Community." In *Reimagining Social Movements: From Collectives to Individuals*, (eds.) Antimo L. Farro and Henri Lustiger-Thaler, pp. 232–249. Surrey: Ashgate.

Douglas, Clifford H. 1920. *Economic Democracy*. San Diego: Harcourt, Brace and Howe.

Drakakis, Helena. 2005. "Interview with P.V. Rajagopal." *Ahmisa Nonviolence*.

Drakakis, Helena, and Simon Williams. 2005. "Truth Force: The Land Rights Movement in India." *Ahimsa Nonviolence* 1 (2): 108–121.

Dworkin, Ronald. 2013. *Justice for Hedgehogs*. Cambridge, MA: Harvard U. Press.

Eisenstadt, S. N. 2009. "Modernity and the Reconstitution of the Political." In *The Modern Prince and the Modern Sage: Transforming Power and Freedom*, (ed.) Ananta Kumar Giri. Delhi: Sage Publications.

Eriksen, Thoams Hylland, and Elisabeth Schober (eds.). 2017. *Knowledge and Power in an Overheated World*. University of Oslo: Department of Social Anthropology.

Escobar, Arturo. 1995. *Encountering Development: The Making and the Remaking of the Third World*. Princeton: Princeton U. Press.

Escobar, Arturo. 2008. *Territories of Difference: Place, Movements, Life, Redes*. Durham: Duke U. Press.

———. 2018. *Designing the Pluriverse: Radical Interdependence, Autonomy, and the Making of Worlds*. Durham, NC: Duke U. Press.

———. 2019. "Civilizational Transitions." In *Pluriverse: A Post-Development Dictionary*, (eds.) Ashish Kothari, Ariel Salleh, Arturo Escobar, Federico Demara and Alberto Acosta, pp. 121–123. Delhi: Tulika Books and Authors Upfront.

———. 2020. *Pluriversal Politics: The Real and the Possible*. Durham, NC: Duke U. Press.

———. 2021. "Anthropocene, Terricide, and the Onto-Epistemic Unframeability of Climate Change." *Article in Great Transition Network*. January 5.

Falk, Richard. 2004. *Declining World Politics: America's Imperial Geopolitics*. London: Routledge.

Fang Fang, and Michael Berry. 2020. *Wuhan Diary: Dispatches from a Quarantined City*. New York: Harper Collins.

Fanon, Frantz. 2004. *The Wretched of the Earth*. New York: Grove Press.

Forst, Rainer. 2017. *Normativity and Power: Analyzing Social Orders of Justification*. New York: Oxford U. Press.

Foucault, Michel. 1984."What Is Enlightenment?" In *Foucault Reader*. London: Penguin.

———. 2005. *Hermeneutics of the Subject*. New York: Palgrave Macmillan.

Francis, Pope. 2015. *Praise Be to You: Laudato 'Si': On Care for Our Common Home*. San Francisco: Ignatius.

———. 2020. *Fratelli Tutti: On Fraternity and Social Friendship*. Vatican: Pope Francis's Encyclical.

Frankl, Victor. 1967. *Psychotherapy and Existentialism: Selected Papers on Logotherapy*. Harmondsworth: Penguin.

Friedman, Maurice. 1965. Chapter 1. "Introductory Essay". Martin Buber, *The Knowledge of Man: A Philosophy of the Interhuman*. New York and Evanston: Harper Torch Books. Quoted in Collins 2018.

Fuad, Zainul. 2007. *Religious Pluralism in Indonesia: Muslim-Christian Discourse*. PhD Theses. University of Hamburg.

Gaewell, Malinn. 2006. *Activist Entrepreneurship: Attacking Norms and Articulating Disclosive Stories*. Stockholm: Stockholm University.

Gandhi, Mohandas Karamchand. 1909. *Hind Swaraj*. Ahmedabad: Navjivan Publishing House.

Garcia Nogales, Mila. 2020. Ecofeminismo Más allá de Greta Thunberg: voces de mujeres indígenas en la lucha medioambiental, Público, Madrid, June 26.

Gautam, Richa, and Anju Singh. 2010. "Corporate Social Responsibility Practice in India: A Study of Top 500 Companies." *Global Business and Management Research* 2 (1): 41–56.

Geertz, Clifford. 1960. *The Religion of Java*. Chicago: University of Chicago Press.

Gessen, Mesha. 2020. "Judith Butler Wants Us to Reshape our Rage." *The New Yorker*. February 9.

Gilbert, Jérémie. 2018. *Natural Resources and Human Rights*. Oxford: Oxford University Press.

Giri, Ananta Kumar. 2002a. "Moral Commitments and Transformation of Politics: Gandhi, Kant and Beyond." In idem, *Conversations and Transformations: Towards a New Ethics of Self and Society*. Lanham, MD: Lexington Books.

———. 2002b. *Building in the Margins of Shacks: The Vision and Practice of Habitat for Humanity*. Hyderabad: Orient Longman.

———. 2006a. "Creative Social Research: Rethinking Theories and Methods and the Calling of an Ontological Epistemology of Participation." *Dialectical Anthropology* 30 (3–4): 227–271.

———. 2006b. "Cosmopolitanism and Beyond." *Development and Change* 37 (6): 1277–1292.

———. 2012. *Sociology and Beyond: Windows and Horizons*. Jaipur: Rawat Publications.

———. 2013. *Knowledge and Human Liberation: Towards Planetary Realizations*. London: Anthem Press.

——— (ed.). 2015. *New Horizons of Human Development*. Delhi: Studera Press.

———. 2016. "Non-Violence in Relations and Non-Injury in Modes of Thinking: Transforming the Subjective and the Objective and the Calling of Transpositional Subjectobjectivity." New Heaven: Lecture presented at Yale University.

Giri, Ananta Kumar. ed. 2017. *Pathways of Creative Research: Towards a Festival of Dialogues*. Delhi: Primus.

———. 2018a. "With and Beyond Epistemologies of the South: Ontological Epistemology of Participation, Multi-Topial Hermeneutics and Contemporary Challenges of Planetary Realizations." Madras Institute of Development Studies: Working Paper.

——— (ed.). 2018b. *Social Theory and Asian Dialogues: Cultivating Planetary Conversations*. Shanghai and Singapore: Palgrave Macmillan.

———. 2018c. *Practical Spirituality and Human Development: Transformation of Religions and Societies*. Sanghai and Sinagapore: Palgrave Macmillan.

———. (ed.). 2018d. *Beyond Cosmopolitanism: Towards Planetary Transformations*. Shanghai and Singapore: Palgrave Macmillan.

———. 2019a. "Cultivating New Movements and Circles of Meaning Generation: Upholding Our World, Regenerating Our Earth and the Calling of a Planetary *Lokasamgraha.*" *Journal of Human Values* 26 (2): 146–166.

———. 2019b. *Weaving New Hats: Our Half Birthdays*. Delhi: Studera Press.

———. 2020a. "Writing Post-Pandemic Life Worlds and Living Words: Ethics, Politics and Spirituality and Alternative Planetary Futures." Paper presented at the International Webinar, "Writing Post-Pandemic Life World: Society, Cultural Materialities and Practices." Department of History, Ravenshaw University, Cuttack, July 22.

———. 2020b. "Trans-Religious Dialogues, Multi-Topial Hermeneutics and Global Responsibility." Paper presented at the Webinar on Science and Religion, Lady Keane College, Shillong, September.

———. 2021a. "Gardens of God." In *Pragmatism, Spirituality and Society: New Pathways of Consciousness, Freedom and Solidarity*, (ed.) Ananta Kumar Giri. New York: Palgrave Macmillan.

———. 2021b. "Cultural Understanding: Multi-*topial* Hermeneutics, Planetary Conversations and Dialogues with Confucianism and Vedanta." *International Communication of Chinese Culture* 8 (1): 83–102.

———. 2021c. "Evolutionity and the Calling of Evolutionary Suffering and Evolutionary Flourishing: Dialogues among Epochs and the Calling of Planetary Realizations." *International Journal of Philosophy*. Special Issue on *Evolutionity*.

———. 2022a. *Alphabets of Creation: Taking God to Bed: A Book of Poems*. Delhi: Authors Press.

———. 2022b. "Chambal Surrender: Remembrance and Resolution on the 50th Year." *Ahimsa* 15 (2): 9–13, May–August.

———. 2023. *Social Healing*. London: Routledge.

Graeber, David. 2011. *Debt: The First 5, 000 Years*. Brooklyn, New York: Melville House.

———. 2013. *The Democracy Project: A History, a Crises, a Movement*. New York: Penguin Random House.

Gramsci, A. 1992. *Prison Notebooks*. New York: Columbia University Press.

Grzybowski, Candido. 2019. "Biocivilization." In *Pluriverse: A Post-Development Dictionary*, (eds.) Ashish Kothari, Ariel Salleh, Arturo Escobar, Federico Demara and Alberto Acosta. Delhi: Tulika Books and Authors Upfront.

Guha, Ramachandra. 2015. "Ramachandra Guha in Conversation with John Harriss." *Development and Change* 46 (4): 875–892.

Habermas, Jurgen. 2003. *The Future of Human Nature*. Cambridge: Polity Press.

———. 1990. *Moral Consciousness and Communicative Action*. Cambridge, MA: The MIT Press.

———. 2020. "Never before Has So Much Been Known about What We Don't Know: Interview with Markus Schwering." *Frankfurter Rundschau*. April 11.

Han, Sang Jin. 2012. "Divided Nation, Unification and Transitional Justice: Why Do We Need a Communicative Approach?" In *Divided Nations and Transitional Justice: What Germany, Japan, and South Korea Can Teach the World*, (ed.) Han Sang-Jin, pp. 1–15. Boulder, CO: Paradigm Publishers.

Handelman, Sapir. 2010. "The Minds of Peace Experiment: A Simulation of a Potential Palestinian-Israeli Public Assembly." *International Negotiation: A Journal of Theory and Practice*, 15: 511–528.

———. 2012a. "Two Complementary Settings of Peacemaking Diplomacy: Political-Elite Diplomacy and Public-Diplomacy." *Diplomacy & Statecraft* 23: 162–178.

———. 2012b. "The Minds of Peace Experiment: A Laboratory for People-to-People Diplomacy." *Israel Affairs* 18 (1): 1–11.

———. 2016. "Peacemaking Contractualism: A Peacemaking Approach to Cope with Difficult Situations of Intractable Conflict." *Global Change, Peace & Security* 28 (1): 123–144.

Handelman, Sapir, and Fredric Pearson. 2014. "Peacemaking in Intractable Conflict: A Contractualist Approach." *International Negotiation* 19: 1–34.

Hans, Asha, Kalpana Kannabiran, Manoranjan Mohanty, and Pusphendra (eds.). 2021. *Migration, Workers and Fundamental Freedoms: Pandemic Vulnerabilities and States of Exception in India*. London and New York: Routledge.

Harari, Yuval Noah Harari. 2018. *21 Lessons for the 21st Century*. New York: Random House.

Haraway, Donna J. 2006. *When Species Meet*. Durham, NC: Duke U. Press.

———. 2016. *Staying with the Trouble: Making Kin in the Chithuluscence*. Durham, NC: Duke U. Press.

Hart, Oliver, and Luigi Zingales. 2017. "Companies Should Maximize Shareholder Welfare Not Market Value." *Journal of Law, Finance, and Accounting* 2 (2): 247–274.

Hartz, Richard. 2014. *The Clasp of Civilizations: Globalization and Religion in a Multicultural World*. Delhi: D.K. Printworld.

Harvey, David. 1989. *The Condition of Postmodernity: An Inquiry into Conditions of Cultural Change*. Cambridge, MA: Blackwell.

Haynes, Jeffrey. 2016. *The United Nations Alliance of Civilizations' Ability to Improve Relations between Christians and Muslims Has Been Limited*. London: London School of Economics.

———. 2017. "The United Nations Alliance of Civilizations and Global South." *Globalizations* 14 (7): 1125–1139.

———. 2018a. "The United Nations Alliance of Civilizations and Interfaith Dialogue: What Is It Good For?" *The Review of Faith and International Affairs* 16 (3): 48–60.

———. 2018b. *The United Nations Alliance of Civilizations (UNAOC) and the Pursuit of Global Justice: Overcoming Western Versus Muslim Conflict and the Creation of a Just World Order*. Chicago: Merlin Press.

———. 2021. "Muslims at the United Nations: Ethical and Political Issues." Paper presented at the International Webinar on "Rethinking and Transforming Public Policy: Ethics, Aesthetics and Responsibility." KIIT Law School and Vishwaneedam Center for Asian Blossoming, May 7.

Heller, Agnes. 1987. *Beyond Justice*. Cambridge, MA: Basil Blackwell.

Higgins, Polly. 2010. *Eradicating Ecocide*. London: Shepherd-Walwyn2nd.

Hogan, John. 2006. "Interview with Rajagopal." *Ahimsa Nonviolence* 11 (1): 57–66.

Honneth, Axel. 2007. *Disrespect: Normative Foundations of Critical Theory*. Cambridge, UK: Polity Press.

Horn, Eva, and Hannes Bergthaller. 2020. *The Anthropocene: Key Issues for the Humanities*. London: Routledge.

Hornborg, Alf. 2017. "How to Turn an Ocean Liner: A Proposal for Voluntary Degrowth by Redesigning Money for Sustainability, Justice and Resilience." *Journal of Political Ecology* 24 (1): 623–632.

———. 2019. *Nature, Society, and Justice in the Anthropocene*. Cambridge: Cambridge U. Press.

Horton, Richard. 2020. *The COVID-19 Catastrophe: What's Gone Wrong and How to Stop It Happening Again*. Cambridge, UK: Polity Press.

Howard, Sarah et al. 2019. "Perspectives on Bioregional Urbanism: Transformative Harmony with Living Systems." In *Transformative Harmony*, (ed.) Ananta Kumar Giri, pp. 317–358. Delhi: Studera Press.

Howard, Sarah with Ninian R. Stein, and Stephen Bissonette. 2019. "Perspective on Bioregional Urbanism: Transformative Harmony with Living Systems." In *Transformative Harmony*, (ed.) Ananta Kumar Giri, pp. 317–358. Delhi: Studera Press.

Huntington, Samuel. 1996. *The Clash of Civilizations and Remaking of World Order*. New York: Simon & Schuster.

———. 2006 [1968]. *Political Order in Changing Societies*. New Haven, CT: Yale University Press.

Iglesias, Pablo. 2015. "Understanding Podemos." *New Left Review* 93 (May–June): 7–22.

Ingold, Tim. 2019. "Art and Anthropology for a Sustainable World." *Journal of Royal Anthropological Institute* 25 (4): 659–674.

Jahanbegloo, Ramin. 2020. *The Courage to Exist: A Philosophy of Life and Death in the Age of Coronavirus*. Hyderabad: Orient Blackswan.

Joseph, Amita, Karandeep Bhagat, Subhash Mittal, Dheeraj, Rohan Preece, and Deepti Menon. 2018. "Will National Guidelines Provide the Much Needed Boost to Business and Human Rights?" In *Status of Corporate Responsibility in India, 2018: Do Business Respect Human Rights?*

Jung, Hwa Yol. 1999. "Postmodernity, Eurocentrism, and the Future of Political Philosophy." In *Border Crossings: Toward a Comparative Political Theory*, (ed.) Fred R. Dallmayr. Lanham, MD: Lexington Books.

Karve, Irawati. 1991 [1969]. *Yuganta: The End of an Epoch*. Hyderabad: Disha Books.

Kelman, Herbert. 2007. "Social-Psychological Dimensions of International Conflict." In *Peacemaking in International Conflict: Methods & Techniques* (revised edition), (ed.) I. William Zartman. Washington, DC: US Institute of Peace.

Khan, Wahiduddin. 2009. *The Prophet of Peace: Teachings of the Prophet Muhammad*. Delhi: Goodword.

———. 2016. *Leading a Spiritual Life*. Delhi: Goodword.

Khairuddin, N., Beverly Yong, and T. K. Sabapathy (eds.). 2013. *Reactions, New Critical Studies: Narratives in Malaysian Art* (Vol. 2). Kuala Lumpur: Rogue Art.

Khatami et al. 2001. *Dialogues Among Civilizations*. Retrieved from the web.

Kim, Dae-jung. 2012. "Power of Dialogue for Peace." In *Divided Nations and Transitional Justice: What Germany, Japan, and South Korea Can Teach the World*, (ed.) Han Sang-Jin, pp. 45–50. Boulder, CO: Paradigm Publishers.

King, Martin Luther Jr. 1967. *Where Do We Go from Here: Chaos or Community?* Boston: Beacon Press.

Knitter, Paul. 1995. *One Earth, Many Religions: Multi-Faith Dialogue and Global Responsibility*. New York: Orbis Books.

Kögler, Hans-Herbert. 2007. "Roots of Recognition: Cultural Identity and the Ethos of Hermeneutic Dialogue." In *Proceedings of the International Wittgenstein Symposium*. Kirchberg, Austria: Ontos Publishing House, May.

———. 2014. "Empathy, Dialogue, Critique: How should We *understand* (Inter-)cultural Violence?" In *The Agon of Interpretations: Towards a Critical Intercultural Hermeneutics*, (ed.) Ming Xie, pp. 275–301. Toronto: University of Toronto Press.

Kolb, Felix. 2005. "The Impact of Transnational Protest on Social Movement Organization: Mass Media and the Making of ATTAC Germany." In *Transnational Protest and Global Activism*, (eds.) Donatella della Porta and Sidney Tarrow. Lanham, MD: Rowman and Littlefield.

Koli, Archana, and Rutvi Mehta. 2020. "Corporate Social Responsibility Practices in the Time of COVID-19: A Study of India's BFSI Sector." www.downtoearth.org.in/blog/governance/corporate-social-responsibility-practices-in-the-times-of-covid-19-a-study-of-india-s-bfsi-sector-74583. Last accessed 9 June, 2021.

Koster, Emlyn. 2019. "The Anthropocene as Our Conscience." In *Designing for Empathy, Perspectives on the Museum Experience*, (ed.) E. Gokcigdem, pp. 344–361. Lanham, MD: Rowman and Littlefield.

Kothari, Ashish, Ariel Salleh, Arturo Escobar, Federico Demara, and Alberto Acosta (eds.). 2019. *Pluriverse: A Post-Development Dictionary*. Delhi: Tulika Books and Authors Upfront.

Kriesberg, Louis. 1993. "Intractable Conflicts." *Peace Review* 5 (4): 417–421.

Kung, Hans. 1996. *Global Responsibility: In Search of a New World Ethic*. New York: Continuum.

Laclau, Ernesto, and Chantal Mouffe. 1985. *Hegemony and Socialist Strategy*. London: Verso.

Lakhani, Nina. 2020. *Who Killed Berta Carcerers? Dams, Death Squads, and an Indigenous Defender's Battle for the Planet*. London: Verso.

Lambek, Michael (ed.). 2010. *Ordinary Ethics: Anthropology, Language, and Action*. New York: Fordham Press.

Landertinger, Laura. 2008. "Brazil's Landless Worker's Movement (MST)." York University, Undergraduate Level Baptista Essay Prize.

Latour, Bruno. 2005. *Reassembling the Social: An Introduction to Actor-Network Theory*. Oxford: Oxford U. Press.

———. 2017. *Facing Gaia: New Lectures on the New Climate Regime*. Cambridge: Polity Press.

———. 2018. *Down to Earth*. Cambridge: Polity Press.

———. 2020. "Is This a Dress Rehearsal?" *In the Moment, Critical Inquiry*. March 26.

Lehmann, Karsten. 2017. *Religious NGOs in International Relations*. London: Routledge.

Lynch, Cecilia. 2012. "Religious Humanitarianism in a Neoliberal Age." *The Religion Factor Blog*, September 12. http://religionfactor.net/2012/09/12/religious-humanitarianism-in-a-neoliberal-age/ Last accessed 29 May, 2013.

Mac Bride Commission. 2003. *Many Voices, One World*. London: Rowman and Littlefield.

Macintyre, Alasdair. 1999. *Dependent Rational Animals: Why Human Beings Need the Virtues*. London: Duckworth.

Maffletone, Sebastiano, and Aakash Singh Rathore (eds.). 2012. *Global Justice: Critical Perspectives*. London: Routledge.

Mander, Harsh. 2020. *Locking Down the Poor: The Pandemic and India's Moral Center*. New Delhi: Speaking Tiger.

Marchetti, Raffaele. 2016. *Global Strategic Engagement*. Lanham, MD: Lexington Books.

Maritain, J. 1971 [1944]. *The Rights of Man and Natural Law*. Sheffield: Gordian Press.

Marotha, Vince. 2009. "Intercultural Hermeneutics and the Cross-Cultural Subject." *Journal of Intercultural Studies* 30 (3): 267–284.

Marshalls, Chris. 2012. *Divine Justice as Restorative Justice*. Victorian University of Wellington: Center for Christian Ethics.

Mbembe, Achille. 2020. "The Universal Right to Breathe." *In the Moment, Critical Inquiry*. April 13.

Melucci, Alberto. 1996. *The Playing Self: Person and Meaning in Planetary Society*. Cambridge: Cambridge U. Press.

Mestres, Laria, and Edward Saleri i Lecha. 2006. "Spain and Turkey: A Long-Standing Alliance in a Turbulent Context?" *Insight Turkey* 8 (2): 117–126.

Mitchell, George. 1999. *Making Peace*. New York, NY: Alfred A. Knopf.

Mohanty, Bindu. 2019. "Climate Change: A Ray of Hope." In *Fridays for the Future*, (ed.) Eugenia Rosca. Scholars' Press.

Mohanty, J. N. 2000. *Self and Other: Philosophical Essays*. Delhi: Oxford University Press.

Mohanty, Manoranjan et al. 2015. "A New Discourse for a Just World: An Introduction." In *Building a Just World: Essays in Honor of Muchukund Dubey*, (eds.) Manoranjan Mohanty, Vinod C. Khanna and Biswajit Dhar, pp. 1–18. Delhi: Orient Blackswan.

———. 2016. "Towards 2030: Global Stirrings from the Himalayan Sphere." In *Exploring the Anthropocene: Towards the Year 2020*, (eds.) Richard Falk and Manoranjan Mohanty, pp. 7–18. Hyderabad: Orient Blackswan.

———. 2021. "Migrant Labour on Center Stage: But Politics Fails Them." In *Migration, Workers and Fundamental Freedoms: Pandemic Vulnerabilities and States of Exception in India*, (eds.) Asha Hans, Kalpana Kannabiran, Manoranjan Mohanty and Pusphendra. 2021. London and New York: Routledge.

Momaday, N. Scot. 2020. "In the Time of Plague." *New York Times*. May 22.

Morgan, Jamie, and Heikki Patomaki. 2021. "Planetary Good Governance after the Paris Agreement: The Case for a Global Greenhouse Tax." *Journal of Environment Management* 292: 1–7.

Murthy, S. V. Ramana. 2004. *The Enchanted Lake: Yaksha Yudhisthira Samvada*. Pune: Rajakiya Sanskrit Sansthan.

Mythen, Gabe. 2018. "Exploring the Theory of Metamorphosis: In Dialogue with Ulrich Beck." *Theory, Culture and Society* 35 (7–8): 173–188.

Nancy, Jean-Luc. 2007. *The Creation of the World or Globalization*. Stony Brook: State University of New York Press.

Ndlovu-Gatsheni, Sabelo J. 2018. *Epistemic Freedom in Africa: Deprovincialization and Decolonization*. London: Routledge.

Norberg-Hodge, Helena. 1991. *Ancient Futures: Lessons from Ladakh in a Globalizing World*. London: Rider.

Nussbaum, Martha. 1996. *Poetic Justice*. Boston: Beacon Press.

———. 2006. *Frontiers of Justice: Disability, Nationality, Species Membership*. Cambridge, MA: Harvard U. Press.

Ocalan, Abdullah. 2013. *Sociology of Freedom: Manifesto for a Democratic Civilization*. Vol. 111. Oakland, CA: Kairos Books and PM Press.

Padel, Felix, Ajay Dandekar, and Jeemol Unni. 2013. *Ecology Economy: Quest for a Socially Informed Connection*. Hyderabad: Orient Blackswan.

Padel, Felix, and Samarendra Das. 2020 [2010]. *Out of This Earth: East India Adivasis and the Aluminium Cartel*. Hyderabad: Orient Blackswan.

País, El. 2019. "Socialists Win Repeat Spanish Election, Vox Becomes Third-Biggest Force in Congress." *EL PAÍS*, November 11. english.elpais.com/elpais/2019/11/10/inenglish/1573407794_574125.html

———. 2020. "Pedro Sánchez Voted Back in as Spanish Prime Minister by Congress." *EL PAÍS*, January 7. english.elpais.com/elpais/2020/01/07/inenglish/1578391109_970993.html

Panikkar, Raimon. 2008. "Introduction." *Mutual Fecundation of Culture* by Varghese Manimala.

Paranjape, Makarand. 2008. "Indian Notions of Responsibility: Self, Society and the World." In Sizoo: 81–96.

Patomaki, Heikki. 2000. "The Tobin Tax: A New Phase in the Politics for Globalization?" *Theory, Culture and Society* 17 (4): 77–91.

———. 2001. *Democratising Globalisation: The Leverage of the Tobin Tax*. London: Zed Books.

———. 2013. *The Great Eurozone Disaster*. London: Zed Books.

———. 2017. *Disintegrative Tendencies in Global Political Economy: Exits and Conflicts*. London: Routledge.

———. 2019. "Emancipation from Violence through Global Law and Institutions: A Post-Deutschean Perspective." In *Pacificism's Appeal*, (eds.) Jorg Kustermans, Tom Sauer, Dominiek Lootens and Barbara Segerent. New York: Palgrave Macmillan.

Perlas, Nicanor. 1997. *Associative Economics: Responding to the Challenge of Elite Globalization*. Quezon City, Philippines: Center for Alternative Developmnet Initiatives.

———. 2003. *Shaping Globalization: Civil Society, Cultural Power and Threefolding*. Quezon City, Philippines: New Society Publishers.

Piketty, Thomas. 2013. *Capital in 21st Century*. Cambridge, MA: Harvard University Press.

———. 2020a. *Capital and Ideology*. Cambridge, MA: Harvard U. Press.

———. 2020b. "Interview with Thomas Piketty ." *Capital and Ideology. Harvard Gazette*.

Pleyers, Geoffrey. 2010. *Alter-Globalization: Becoming Actors in a Global Age*. Cambridge: Polity Press.

Podemos. 2019. *Programa de Podemos*. https://podemos.info/wp-content/uploads/2019/10/Podemos_programa_generales_10N.pdf

Pogge, Thomas. 2001. "Priorities of Global Justice." *Metaphilosophy* 32 (1).

———. 2002. *World Poverty and Human Rights*. Cambridge: Polity Press.

Prahalad, C. K. 2004. *The Fortune at the Bottom of the Pyramid*. Philadelphia: Wharton School Publishing.

Puchala, Donald J., KatieVerlin Laatikainen, and Roger A. Coate. 2007. *United Nations Politics: International Organization in a Divided World*. Upper Saddle River, NJ: Prentice Hall.

Quarles van Ufford, Philip, and Ananta Kumar Giri (eds.). 2003. *A Moral Critique of Development: In Search of Global Responsibilities*. London: Routledge.

Rajagopal, P. V. 2005. "Solidarity with the Forgotten: North-South Solidarity for the Future of Voluntary Activism." *Ahimsa Nonviolence* 1 (5): 465.

Rajavelu, K., and Stanley Joseph. 2019. "Justice: A Distant Dream in Thoothukudi." In *Status of Corporate India, 2019: Is Human Rights in Business Limited to Rhetoric?*

Ramesh, Jairam, and Muhammad Ali Khan. 2015. *Legislating for Justice: The Making of the 2013 Land Acquisition Act*. Delhi: Oxford U. Press.

Rao, Sailesh. 2016a [2011]. *Carbon Dharma: The Occupation of Butterflies*. Phoenix: Climate Healers Publications.

———. 2016b. *Carbon Yoga: The Vegan Metamorphosis*. Phoenix, AZ: Climate Healers Publications.

Rathore, Aakash Singh. 2012. "The Romance of Global Justice: Sen's Deparochialization and the Quandary of Dalit Marxism." In *Global Justice: Critical Perspectives*, (eds.) Sebastiano Maffletone and Aakash Singh Rathore. London: Routledge.

Rawls, John. 1971. *A Theory of Justice*. Cambridge, MA: Harvard U. Press.

———. 2001. *Justice as Fairness: A Restatement*. Cambridge, MA: Harvard University Press.

Reardon, Jenny. 2020. "V Is for Veracity." *Items*. New York: Newsletter of Social Science Research Council.

Reid, Herbert, and Betsy Taylor. 2010. *Recovering the Commons: Democracy, Place and Global Justice*. Urbana Champaign: University of Illinois Press.

Reubke, Karl-Julius. 2020. *Struggles for Peace and Justice: India, Ekta Parishad and the Globalization of Solidarity*. Delhi: Studera Press.

Ricouer, Paul. 2000. *Just*. Chicago: University of Chicago Press.

Rifkin, Jeremy. 2010. *The Empathic Civilization: The Race to Global Consciousness in a World in Crises*. London: Jeremy P. Tarcher, Inc.

Robinson, Andy. 2011. "Spain's Indignados Take the Square." *Nation*. 155 June 8, 2011

Rodrigue, Barry, Leonid Grinnin, and Andrey Korotayev. 2017. *The Ways That Big History Works: Cosmos, Life, Society and Our Future*. Delhi: Primus Books.

Rodriguez, Maria Natalia P. 2019. *Social Safeguards and Equity in the Provisions of Payment for Environmental Services in the Paris Agreement*. PhD Thesis, University of Geneva, Geneva.

Ross, Ulrich. 2018. "Cosmopolitanism, Spirituality and Social Action: Mahatma Gandhi and Rudolf Steiner." In *Beyond Cosmopolitanism*, (ed.) Ananta Kumar Giri. New York: Palgrave Macmillan.

Roy, Arundhati. 2020. "The Pandemic Is a Portal." In idem, *Azadi: Freedom, Fascism, Fiction*. New Delhi: Penguin Random House.

Royal, Te A. C., and Betsan Martin. 2008. "Indigenous Ethics of Responsibility in Aotearoa/New Zealand: Harmony with the Earth and Relational Ethics." In *Responsibility and Cultures of the World*, (ed.) E. Sizoo, pp. 47–64. New York: Peter Lang.

Sachs, Jeffrey. 2012. *The Price of Civilization: Reawakening Virtue and Prosperity after the Economic Fall*. London: Vintage.

Sandal, Nukhet, and Jonathan Fox. 2015. *Religion in International Relations Theory*. London: Routledge.

Sandel, Michel. 1982. *Liberalism and the Limits of Justice*. Cambridge: Cambridge U. Press.

Sen, Amartya. 1993. "Positional Objectivity." *Philosophy and Public Affairs* 22 (2): 126–145.

———. 1999. *Development as Freedom*. New York: Alfred A. Knof.

———. 2002. "Justice across Borders." In *Global Justice and Transnational Politics: Essays on the Moral and Political Challenges of Globalization*, (eds.) P. D. Grieff and C. P. Cronin, pp. 37–51. Cambridge, MA: The MIT Press.

———. 2008. "Foreword." In *To Uphold the World: The Message of Ashoka and Kautilya for the 21st Century*, (ed.) Bruce Rich. Delhi: Penguin.

———. 2009. *The Idea of Justice*. London: Allen Lane.

———. 2012. "Global Justice." In *Global Justice: Critical Perspectives*, (eds.) Sebastiano Maffletone and Aakash Singh Rathore. London: Routledge.

Sharma, Subhash. 2007. *New Mantras in Corporate Corridors: From Ancient Roots to Global Routes*. New Delhi: New Age Publishers.

———. 2012. *New Earth Sastra: Towards Holistic Development and Management*. Bangalore: IBA Publications.

———. 2020. "Corporate Spiritual Responsibility: Towards a New Paradigm for Corporate Social Responsibility." *3D: IBA Journal of Management and Leadership* 11 (2).

Simón, Pablo. 2020. "The Multiple Spanish Elections of April and May 2019: The Impact of Territorial and Left-Right Polarisation." *South European Society and Politics*. DOI: 10.1080/13608746.2020.1756612

Singh, Rustam. 2011. *"Weeping" and Other Essays on Being and Writing*. Jaipur: Pratilipi Books.

Sizoo, Edith. 2000. *What Words Do Not Say: Perspectives for Reducing Intercultural Misunderstandings: The Singular Experience of Translating the Platform of the Alliance for a Responsible and United World*. Paris: Editions Charles Leopold Mayer.

——— (ed.). 2008. *Responsibility and Cultures of the World: Dialogue Around a Collective Challenge*. Bruxells et al. Peter Lang.

Smith, Jackie et al. 2014. *Global Democracy and the World Social Forum*. London: Routledge.

Smith, Linda Tuhiwai. 2012 [1999]. *Decolonizing Methodologies: Research and Indigenous Peoples*. London: Zed Books.

Sneller, Rico. 2020. *Perspectives on Synchronicity, Inspiration, and the Soul*. New Castle Upon Tyne: Cambridge Scholars Press.

Sola, Jorge, and César Rendueles. 2018. "Podemos, the Upheaval of Spanish Politics and the Challenge of Populism." *Journal of Contemporary European Studies* 26 (1): 99–116. DOI: 10.1080/14782804.2017.1304899

Sorokin, Pitrim A. 1985. *Society, Culture and Personality: Their Structure and Dynamics*. NY: Harper & Brothers.

Sparks, Allister. 1994. *Tomorrow Is Another Country: The Inside Story of South Africa's Negotiated Revolution*. Sandton, South Africa: Struik Book Distributors.

Sri Aurobindo. 1993 [1950]. *Savitri*. Pondicherry: Sri Aurobindo Ashram.

Sristhi and Tavleen Singh. 2020. "Why This Is Not CSR: A Study of 5 Major Corporates." www.downtoearth.org.in/blog/governance/why-this-is-not-csr-a-study-of-5-major-corporates-74587

Steiner, Rudolf. 1985. *Renewal of the Social Organism*. Goetheanum, Switzerland: Anthroposophic Press.

Strathern, Marilyn. 2021. "Regeneration and Its Hazards." Paper presented at the Swadhyaya Sahachakra Circle, Online Meeting, November 7.

Strydom, Piet. Forthcoming. Foreword to Corporate Spiritual Responsibility eds Ananta Kumar Giri & Subhash Sharma.

Strydom, Piet. 1999. "The Civilization of the Gene: Biotechnology Risk Framed in the Responsibility Discourse." In *Nature, Risk and Responsibility: Discourse of Biotechnology*, pp. 21–36. London: Palgrave Macmillan.

———. 2000. *Discourse and Knowledge: The Making of Enlightenment Sociology*. Liverpool: Liverpool University Press.

———. 2002. *Risk, Environment and Society*. Berkshire, UK: Open University Press.

———. 2009. *New Horizons of Critical Theory: Triple Contingency and Collective Learning*. Delhi: Shipra.

———. 2015. "Cognitive Fluidity and Climate Change: A Critical Social-Theoretical Approach to the Current Challenge." *European Journal of Social Theory* 18 (3): 236–256.

———. 2018. "On the Age of Responsibility: The Responsibility Discourse and the Prospects of a Responsible Society." In *Quels dendemairs pour la responsibilitie? Perspectives multidisciplinaires*, (eds.) Allison Marchildon and Andre Duhamel. Montreal: Nota Bene.

Forthcoming. Foreword to *Corporate Spiritual Responsibility*, eds. Ananta Kumar Giri, Subhash Sharma and Ramana AV Achyarulu.

Sundara Rajan, R. 1998. *Beyond the Crisis of European Sciences: New Beginnings*. Shimla: Indian Institute of Advanced Studies.

Swaminathan, M. S. 2011. *In Search of Biohappiness*. Singapore: World Scientific.

Swaminathan, M. S., and Daisaku Ikeda. 2005. *Revolutions: To Green the Environment, to Grow the Human Heart*. Chennai: East-West Books.

Tagore, Rabindranath. 1917. "Introduction." *The Web of Indian Life* by Sister Nivedita. Kolkata: Advaita Ashram.

Tanabe, Akio. 2020. "Politics of Relationship in the Anthropocene: A Search of Well-Being of Human Co-Becomings." Paper presented in the International Webinar on "Writing Post-Pandemic Life World: Society, Cultural Materialities and Practices." Department of History, Ravenshaw University, Cuttack, July 17.

Tarrow, Sidney. 1998. *Power in Movement: Social Movements and Contentious Politics*. Cambridge: Cambridge University Press.

Taylor, Betsy. 2011. "Civil Society, Social Movements and Alternative Development: Implications of Giri's Notion of Knowledge." *Sociological Bulletin* 60 (1).

Thomas, Scott. 2005. *The Global Transformation of Religion and the Transformation of International Relations: The Struggle for the Soul of the Twenty-First Century*. New York and Basingstoke, UK: Palgrave Macmillan.

Thoreau, Henry David. 1947. "Walking." In *Portable Thoreau*. New York: Penguin.

Thunberg, Greta. 2019. *No One Is Too Small to Make a Difference*. New York: Penguin.

Torrez, Faustina. 2011. "La via Campesina: Peasant-Led Agrarian Reform and Food Sovereignty." *Development* 54 (1): 49–54.

Unger, Roberto M. 2004. *False Necessity: Anti-Necessitarian Social Theory in the Service of Radical Democracy*. London: Verso.

UNHCR. 2020. *Spain Asylum Applications Q1–4 (1 Jan–31 Dec2019)*. January 25. https://data2.unhcr.org/en/documents/download/73417

van Staveren, Irene. 2020. "The Economic Consequences of the Corona Crisis." Student-Staff Dialogue between Vincenzo D'Egdio & Irene van Staveren. *ISS News* 22 (1): 19.

Varela, A. 2019. *Qué medidas proponen exactamente sobre inmigración PSOE, PP, Ciudadanos, Podemos, Vox y Más País*. Retrieved June 17, 2020, from www.businessinsider.es/inmigracion-proponen-pp-psoe-podemos-ciudadanos-vox-404701

Varughese, John. 2012. *Truth and Subjectivity, Faith and History: Kierkegaard's Insights for Christian Ethics*. Eugene: Wof and Stock.

Vattimo, Gianni. 1999. *Belief*. Cambridge: Polity Press.

Vattimo, Gianni. 2002. *After Christianity*. New York: Columbia U. Press

———. 2011. *A Farewell to Truth*. New York: Columbia U. Press.

Visvanathan, Shiv. 1996. "On Unravelling Rights." *Studies in Humanities and Social Sciences* 2 (2): 109–149.

———. 2017. "The Search for Cognitive Justice." In *Research as Realization: Science, Spirituality and Harmony*, (ed.) Ananta Kumar Giri, pp. 247–256. Delhi: Primus Books.

Weir, Margaret. 2020. "The Pandemic and the Production of Solidarity." In *Items*. New York: Newsletter of Social Science Research Council.

Willis, David Blake. 2021. "Gandhi and Aurobindo in the Age of Corona: Reflections on Transformative Leadership, End Times and the Kali Yuga." Afterword to *Mahatma Gandhi and Sri Aurobindo*, (ed.) Ananta Kumar Giri. London and New York: Routledge.

Wiwa, Ken-Saro. 1995. *A Month and a Day: A Detention Diary*. New York: Penguin.

Wolf, Susan. 2016. "Aesthetic Responsibility." The Amherst Lecture in Philosophy. Lecture 11. https://amherstlecture.org

Wolford, Wendy. 2003. "Producing Community: The MST and Land Reform Settlement in Brazil." *Journal of Agrarian Change* 3 (4): 500-520.

Wolford, Wendy. 2010. *This Land I Own Now: Social Mobilization and the Meanings of Land in Brazil*. Durham: Duke University Press.

Young, Iris Marion. 2011. *Responsibility for Justice*. Oxford: Oxford U. Press.

Yunus, Muhammad. 2008. *Creating a World Without Poverty: Social Business and the Future of Capitalism*. New York: Public Affairs.

———. 2010. *Building Social Business: The New Kind of Capitalism That Serves Humanity's Most Pressing Needs*. New York: Public Affairs.

Zizek, Slavos. 2020. *Covid-19 Shakes the World*. New York and London: OR Books.

Index

For Product Safety Concerns and Information please contact our EU
representative GPSR@taylorandfrancis.com
Taylor & Francis Verlag GmbH, Kaufingerstraße 24, 80331 München, Germany

9 7 8 1 0 3 2 4 4 1 8 6 3